MINI

The True and Secret History of the Making of a Motor Car

Simon Garfield is the author of eleven works of non-fiction including *Mauve*, *The Error World* and *The Nation's Favourite*. His edited diaries from the Mass Observation Archive – *Our Hidden Lives*, *We Are at War* and *Private Battles* – provided unique insights into the Second World War and its aftermath, and his study of AIDS in Britain, *The End of Innocence*, won the Somerset Maugham Prize.

For five years he was driven to school in a neighbour's Mini, one of the early ones with the pull-down plastic cable for a door handle and an overpowering smell of damp.

MINI

The True and Secret History of the Making of a Motor Car

Simon Garfield

faber and faber

First published in 2009
by Faber and Faber Limited
Bloomsbury House
74–77 Great Russell Street
London WC1B 3DA
This paperback edition first published in 2010

Typeset by Faber and Faber Limited
Printed in England by CPI BookMarque, Croydon

A CIP record for this book
is available from the British Library

ISBN 978–0–571–24811–7

10 9 8 7 6 5 4 3 2 1

To everyone who made the car

Contents

Preface

On Monday 3 November 2008, a man arrived at MINI Plant Oxford to make a car. It was 6.15 a.m., still dark, and from all over the surrounding area other men and women began arriving too – on bicycles, in old Vauxhalls and Fords, some in Minis, some in MINIs, most alone, a few in pairs. Ahead of them lay a ten-hour shift. Some of them carried their lunch in small rucksacks on their backs.

That man on his own – that was me. I was there to learn how to build a MINI, a day's training with two other men. The other two would do their best not to show me up, but they were people who thought with their hands. They knew what a subframe was, and DC tooling, and all about newton metres, and in time they could probably build a whole engine, and then a body shell, and eventually a car.

For my part, I could turn on a digital tape recorder.

Making a car was not the task it once was. I had heard stories about Dante's *Inferno*, of oil slicks beneath men's feet, of industrial fatalities. The place I was visiting, at Cowley on the outskirts of Oxford, was the place where British cars had been built for almost a hundred years, and amidst the grease and infernal clatter it concealed the most romantic story. This is where the great motor-car builder William Morris set up shop, where the cars of our parents and grandparents were made. It was where hundreds of thousands of people spent a lifetime welding metal together. And it was where the Mini reinvented itself as the MINI.

Born out of an economic and political crisis, it celebrates its fiftieth birthday during another one; once more it may be the best car for its time. When the first Mini was sold in 1959, no one – not even its egomaniacal designer – guessed what a thing it would become: more than a car. People would talk of it not only as a means of transport but as a lifestyle and an attitude. People would think of the car and smile. How could this be? Why would marketing people claim that more drivers gave their Mini a name than any other car? How would a thin box of metal and wires become a welcome part of a human family?

The Mini of old is not the MINI of today – thank God. Its single greatest achievement – popularising the idea that small-ness and not bulk is desirable – has held fast through half a century of great turbulence in the British car industry, and it may be the only constant. The car has improved beyond measure. It has a new style, a new size, a new price, but it maintains strong links with the past not just through its factory but through its ethos. It is no longer British, but this is not uppermost in the minds of the hundreds of Poles and Turks and Greeks and Ukrainians and Irish who turn up each day to build it.

This book is the story of the making of a car by the people who made it in 1959 and who make it in 2009, some of whom have grand titles and some of whom do not. It is not an account of the Mini's rallying success or of the fans who customise the cars or the cars' celebrity owners. It is not even an account of blowing the bloody doors off. There are already some fine books about these things, but this is an attempt to relate a more intimate story. When I turned up to build the car at the beginning of the recession at the end of 2008 it became clear that alongside the engineering and technological feats of the cars lay a tale of immense human endeavour, and I hoped that perhaps an oral narrative assembly line could reveal new truths. The people I spoke to tell the story as best as they can remember it, and because the story of the Mini old and new means so much to them, they tend to remember it vividly and with great affection. I do wonder whether any other car has ever been so intensely fought for, or so wildly loved.

Cast of characters

(in order of appearance)

PART ONE

Peter Tothill, a production engineer
Roy Davies, a vehicle proving engineer
Alec Issigonis, a car designer
Alex Moulton, a rubber suspension specialist
Donald Stokes, a manager of British Leyland
William Morris, a manufacturing pioneer
Laurence Pomeroy, a motoring expert
Tony Ball, a publicity man
John Cooper, a speed engineer
Ronald 'Steady' Barker, a motoring correspondent
Lord Snowdon, an enthusiast
Eddie Cummings, a safety engineer
Jean Cummings, a trim machinist

PART TWO

Paul Chantry, a former deputy director
Ian Cummings, a process improvement manager
Chris Bond, a deputy union man
Frank Stephenson, a cool designer
Pat Nolan, an engineering manager

Gert Hildebrand, a chief designer
Peter Ustinov, a renaissance man
Elaine Butler, a personal assistant
Donna Green, a quality specialist
Peter Crook, a paint director
Cedric Scroggs, an old-school marketing director
Bernard Moss, a plant convenor
Emma Lowndes, a modern marketing director
Frau Dr Larissa Huisgen, a Munich-based promotional expert
Dr Herbert Diess, a former Oxford plant director
Jeremy Clarkson, a voice of motoring
Mike Colley, an assembly line trainer
Andy Lambert, an assembly line director
Richard Clay, an assembly line trainer
Jim McDowell, an American vice president
Frau Dr Stefanie Ludorf-Ring, an events and corporate
    meetings expert
Giles Smith, a motor car columnist
Oliver Zipse, an outgoing plant director
Dr Jürgen Hedrich, an incoming plant director
Gabrielle Hummelbrunner, an advertising and promotional
    film coordinator

PART ONE

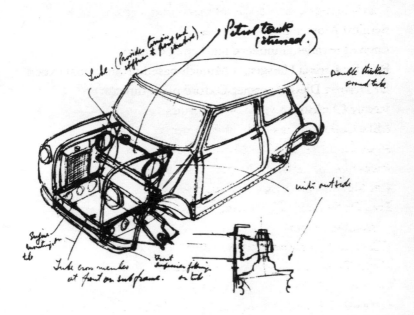

1 'This car is going to leak,' said Jim Percival

Peter Tothill (production engineer)
Towards the end of 1957, Leslie Ford, the chief planning engineer, called me into his office and said, 'There is a new highly secret model coming, and I want you to deal with it from scratch. I will take you over to E Block paintshop and show you where we will build it.'

We entered the south-east corner of the building and there was a vast area of pristine concrete floor.

So where the hell do I start?

Roy Davies (vehicle proving engineer)
Early in 1958, a prototype Mini appeared in the chassis experimental department next door to the drawing office where I worked. It was quickly locked away in a cabin, and access was only allowed to authorised persons. The cabin was guarded by a works policeman.

Peter Tothill
The next task was to get some idea of the car. It was coded ADO15 (Austin Drawing Office, project 15) but already nicknamed Sputnik. I had already seen this strange little vehicle with wheels the size of shirt buttons. It was Prototype No. 1, and was being endurance-tested round and round Chalgrove aerodrome, south-east of Oxford.

Roy Davies

The Mini was called Sputnik because at the time they started doing the testing round Chalgrove, the Russians had just put the Sputnik in the air, an orange thing going round the sky. At night, the Mini going round Chalgrove was an orange thing going round on the ground. The road-proving engineers christened it Sputnik much to Issigonis's disgust. In later years he had a change of memory and said that he christened it, but he didn't.

Alec Issigonis (designer)

At that time, Leonard Lord [chairman of the British Motor Corporation] got me to design a car to beat these bubble cars.

Alex Moulton (suspension engineer)

He'd been called Arrogonis.

Alec Issigonis

Oh yes, I'm Arrogonis!

Leonard Lord gave me completely carte blanche to do what I wanted to do, except one thing. He said, 'Alec, you can do whatever you like, but it's got to have an existing engine that is now in production.'

Lord Stokes (chairman, British Leyland)

Alec Issigonis, as you know, a very great man, a very charming man, one of the most charming men you could possibly meet, but he was also – and this is why he succeeded I suppose – a very *dominant* engineer.

Alec Issigonis

Lord got in the [prototype] with me and I took him round the block at Longbridge as fast as I could – they've got a big circuit at the works – and I almost frightened him I think. We stopped outside his office, and he said, 'Alec, I want it made in a year's time. Get going.'

Never, ever copy the opposition: Alec Issigonis poses at Longbridge

Peter Tothill

So the next thing that happened was we were loaned Prototype No. 3 to strip and rebuild and we had a little compartment built, the cabin, in the experimental department, out of plywood sheets so that we could do it in some secrecy. We were a bit vulnerable from the press because the way the bypass was built, you could see right into the factory from the ramp. Our cabin was built over a pit area so we could get underneath.

There was a small team of us, and we had a week to strip and rebuild this car, so we did that and I made endless notes, problems or potential problems.

Issigonis – I may appear very critical of Issigonis, which I am in a lot of ways – but one of the things he decided, and he convinced the board, was that you didn't need a vast engineering organisation. Now Issigonis was a brilliant innovator, absolutely first-class innovator, but he had no time for the nitty-gritty detail. He just used to say, 'Solve it, don't come to me, just solve it.'

Alec Issigonis

I design cars without any prompting from my employers to suit what they want for sale. I thought I knew better than the market-research people what the public wanted. As is shown in the results.

Roy Davies

The thing that rolled through the Mini was the water leak fiasco.

Issigonis for a long time would not admit that there was a water leak, despite the fact that his Mini had a pair of wellingtons constantly in the boot for him to wear. The basic problem was, if Issi had designed a house, he would have put the tiles on upside down. When he designed the Mini floor, he insisted on having it so that the water went up. Instead of having the floor like that [hands curved upwards] he insisted on it being that way [facing down], so the water just went into the car. The boys at Nuffield Metal Products who make the bodies and the people at the Fisher division who make the bodies all said,

'It's not sealable.' It took Issigonis about eight or nine months to acknowledge that it wasn't sealable.

Peter Tothill
Before we'd even started building I went up to a meeting in Birmingham at Nuffield Metal Products who built the bodies for Morris. My understanding was that the only purpose of the meeting was to establish in detail how they were going to ship the body to Cowley, what would be fitted, what wouldn't be fitted. After the meeting ended, a guy called Jim Percival, who was the engineering director, said, 'Right, I want to talk to you.' I thought, what have I done wrong? He said, 'I want you to talk to your boss Leslie Ford when you get back. I've already told him, but I want you to reinforce it: this car is going to leak.' So I thought, well, why don't you bloody well change it? He said, 'I've had three acrimonious meetings with Issigonis and he will not change the design of the body. He's done it according to him; he's done it for ease of assembly. I've argued with him that it doesn't make it any easier to assemble, but it does guarantee it will leak.' Issigonis insisted on it being like that, so that any water coming off the front wheels went straight [up into the floor], and the pressure of it coming off – it was like coming out of a fire-hose jet. 'When it comes off the wheel it will wash any sealant out and it'll leak. And there's another place in the front wheel arch,' he said, 'where once it's got past the sub-assembly stage you can't get at it to seal it.' He said, 'I want you to impress this on Leslie Ford.' Fine. I did that and Leslie Ford reported it up through the process but didn't get anywhere with Issi.

MINI

Roy Davies

What was interesting, it was conceived really because of the Suez Crisis and so forth, and yet the take-up by the public was initially very low. But at the car factory it made an immediate impression on you. Because [before it] there was the Morris Minor and the Morris Oxford. Before that was the Morris Six, which was a big Morris Oxford with a longer bonnet. And the Wolseley Six was a Morris Six with a different front end on it. We'd always grown this family of cars, but the Mini was almost like dropping a Ford Escort into the place. It was immediately different. The shape instantly gelled. The shape looked right – it was a complete car.

2 'Our family luck was out,' said William Morris

William Morris
Mine was an unexceptional boyhood in many ways, as this plain, unemotional but, for once, factually accurate array of statements will, I hope, prove.

In November 1937, W. M. W. Thomas, managing director of Wolseley Motors Ltd in Ward End, Birmingham, conducted a series of interviews with William Morris (Lord Nuffield) to get some straight information about his early life. A short while later, he sent Morris the following draft for approval.

I was not, as so many imagine, an only child. Actually, I had at one time six brothers and sisters. My birthplace was Worcester, but for all practical purposes we were brought up in Headington, near Oxford.

We were neither very poor nor what is known as rich. The farm, on which I spent my boyhood days, was the home of my mother's family. I did what most young boys would do – rode on horses, helped in the harvest fields and lived a healthy, natural out-of-door life until I was in my teens.

My school days at Cowley were uneventful. As a family, as a matter of fact, we had rather a complex about schools and behaviour therein. It arose from the fact that during his school days my father, who stood six feet at the age of seventeen, had,

Morris Garages triumphant: William Morris with a winner from 1932

in a sense, disgraced himself by knocking down and nearly seriously injuring his form-master . . . Naturally there were consequences, and it was particularly impressed on me by my mother that gentleness and self-restraint were virtues always to be cultivated.

My father was a man who was full of almost unreasoning energy. After his early life at home, but before meeting my mother, he went to North America and, his adventurous spirit taking him into the remoter parts of the continent, he had the distinction of being one of the very few men who were honoured by the title Chief, ritually administered by a tribe of native Indians.

I suppose my innate love of travel and transportation comes from him. For quite a considerable time he drove the Mail Coach Service in newly developed lands. He was a pioneer in the true sense, but like several other pioneers who helped to build this brave modern world, he paid the penalty later with his health.

When I was about sixteen it plainly became apparent that my father's health was failing. Racking attacks of asthma prevented him enjoying a full life of work or even pleasure. Our family luck was out.

Although young, I could see my mother wistfully missing the little luxuries to which, as a girl gently matured and well educated, she had been accustomed. We moved from the farm at Headington to a smaller house in Oxford. It quickly dawned on my young life that for me there would be no leisurely years finishing my education. I must get into harness quickly, and straight away, almost, I developed a habit that has stayed with me ever since – the habit of making money.

In the draft, this was amended by Morris in pencil to read: '– the habit of working.'

In my view, the worst thing that can happen to a young fellow starting life is for him to have sufficient financial security to take the keen edge of endeavour off his outlook.

But anyway, that didn't happen to me.

Morris had been in the vehicle trade since 1892.

At the age of fifteen, with £4 capital, he began repairing and then making bicycles, and the trophies he won racing them in county championships decorated his office. By 1900 he had moved from the garden shed at his parents' house to a shop and

Within sight of my ambition: new premises in 1913

garage in Oxford's High Street, and here he installed the castings for a 1¾-horsepower engine to be fitted to his bikes. The sign above the window read 'WILLIAM R. MORRIS, MOTOR CYCLE ENGINEER', and on the window itself was etched the word 'CELEBRATED'.

I enjoyed making those bicycles. The newly formed mental habit of thinking in terms of making rather than spending money made me look on them as sources of profit as well as pleasure. I realised the keen pleasure a born mechanic feels when a joint brazes smooth and firm, when a built-up, wire-spoke wheel spins true and free in its bearings. I enjoyed racing those machines that I had made, and I was happy in the knowledge that I was helping to maintain our domestic establishment in comfort and occasional luxury.

Until – and this is the tragedy of my life – my mother's health also showed signs of failing. In spite of the care and good medical attention that I was able to provide from the then small, but growing business I had built up, my mother developed a tubercular lung.

It was then vividly seared on my impressionable mind how more valuable than all this world's riches is bodily health.

I worked in those days to gain means to cure my parents. That is the thread that goes clear through the warp of my life.

It made me work and formed my life round the habit of never needlessly wasting money on unproductive things; though never, I hope, grudging the full and proper use of wealth, which is to improve the lot of mankind.

I was fortunate to find a life partner who shared in these ideals. To her I owe a great deal of my success. We have a clear division of activities. Her sphere is the home; mine the office and the factory. My most casual plans we discuss together in the evening. We have ridden hundreds of miles on tandem bicycles through the length and breadth of the land . . . Many of the plans for the early developments of Morris Garages Ltd and WRM Motors Ltd were talked over while we ate a picnic meal together on these frequent excursions.

They did not have children. He may not have had time: he said he would not rest until every factory worker went to work by car.

By 1906, Morris had set his horizons beyond motorbikes. He hired cars to university people, and repaired those that came in, often getting under the bonnet to see how they worked. He also ran a taxi cab. His new garage at Longwall became known locally as the Oxford Motor Palace, and his showroom selling Triumph, Enfield and Douglas motorbikes expanded into an agency for Humber, Singer, Belsize, Standard, Arrol-Johnston and Hupmobile cars. When the Prince of Wales came to Magdalen he serviced his Daimler Tourer at Longwall, and continued to use the garage after he left the university.

The cars Morris worked on were usually custom-built, heavy, thirsty and beyond the reach of all but the wealthiest. He thought he could make something cheaper, lighter and more reliable.

With care and by exercising the strictest economies we had built up the Garage business of respectable dimensions and were within sight of realising my life's ambition – to build a Morris Car.

He struck a deal with the Coventry firm White and Poppe to supply carburettors, gearboxes and ten-horsepower four-cylinder engines. It was the start of a familiar process: Morris would soon be pressing the car bodies himself, but he would obtain key components from established manufacturers. If they couldn't supply him with the quantities he needed, he would take them over and improve efficiency.

He exhibited plans for his Morris-Oxford Light-Car at the 1912 Motor Show, where interested parties may already have begun to call it the Bullnose (its radiator had an unusual domed steel top), and the literature proclaimed it as a luxury item at one quarter of the price of the cheapest Daimler, and cheaper than a comparative Humber and Singer.

Price: £175 with two-seater torpedo body.
Tax £3. 3. 0.
Dunlop 700 x 80 mm tyres.
Leather upholstery.
Brass mounts.
Cape hood and adjustable windscreens.
Full set of Powell and Hanmer lamps.
50 to 55 miles per gallon.
Speed range 5 to 55 miles.
3-speed gearbox.

10 hp White and Poppe engine with side valves and adjustable tappets and a partially detachable crankcase.

Forced feed lubrication.

H-section front axle of forged steel.

Worm and worm-wheel steering.

Two sets of shoes in rear wheel drums.

One dealer at the motor show was particularly impressed, buying four hundred without seeing the finished product. The dealer, Gordon Stewart, later made his fortune with the sole London distribution contract for all Morris cars between the wars, and Morris had the order book he needed to move production to Cowley.

By 1914, that fateful August, our plans were ready. We could go into production. But beyond a very few cars we had to cease our attempts at manufacture. The disappointment, the change of the whole tempo of the business, accelerated my father's illness. He died in 1916. It was due solely to the encouragement of my wife and mother that I was able to face up to the bitter blow of the War. We [ran] a three-shift, 24-hour system of producing mine sinkers and trench mortars. I was a salaried manager in my own factory – the factory building, strangely enough, being the self-same Hurst's Grammar School where my father had laid-out the form-master. My office was then, and still is to this very day, the old Headmaster's study.

The flow-line assembly process producing munitions was immediately adapted to cars, but the immediate postwar trade boom of 1919–20 was followed by a disastrous slump. Sales at Morris Motors Ltd were 276 in September 1920, seventy-four in January 1921. Morris judged this to be mostly a matter of costs, and he cut the price of his cars by a fifth. Sales immediately improved, and the further reductions he made at the 1921 Motor Show established his company as one of the leading forces in the industry. At the end of that year sales had reached three thousand; in 1922 they were almost seven thousand; by 1925 they stood at fifty-five thousand.

The range expanded swiftly: the Oxford gave way to the Cowley, the bullnose radiator to the flatnose, the two-seater to the four-seater tourer, the three-gear box to four. Morris began founding new companies – MG (Morris Garages) sports cars began in 1923 with a souped-up car winning the London-to-Land's End trial, and a new factory opened at Abingdon to meet demand. In 1925, his own printing press began producing the marketing and technical literature for his cars, and soon The Morris Owner *magazine. He bought Wolseley Motors in 1926, Riley in 1938, and he formed Morris Pressings in the same year, confirmation that in future his cars would have all-steel bodies. He acquired engine, radiator and carburettor companies as the expansion of his empire demanded, and established a network of factories in the Midlands, Wales and the south of England that kept the connecting roads full of Morris commercial vehicles and trucks. The secret was simple: volume production and what, in a later age, came to be known as synergy.

The first Morris Minor had been launched in 1928, Oxford's answer to Birmingham's successful Austin Seven. It sold for*

£125, and was capable of 50 mph and 50 mpg. But the depression brought a shrunken market and further competition (increasingly from Ford's small cars), and the engine on the Minor was modified to enable a dealership sale price of £100. The adverts now promised an astonishing 100 mph and 100 mpg, but both feats were achieved separately, the first at Brooklands with a turbo attached, and the second travelling on a circuit between Birmingham and Coventry at 15 mph. Within a year it accounted for half of UK car production.

William Morris's office in Cowley, his main place of work for fifty years, contained relatively few signs of his unrivalled impact on the British motor industry. There were no pictures or plans of cars, only a few replicas presented as gifts. There were some models of aeroplanes, a framed print of Kipling's poem 'If', three barometers, his cycling medals and trophies, a book on pharmacy and a tin of bicarbonate of soda. The room was painted green; the fields beyond his window were green once too, but had been covered by an unbroken stretch of brick and mortar: buildings housing men producing cars.

It is a sight that my mother never saw. With the encroaching years she became more and more reserved. Hers was the dignity of the Nineteenth Century. She could have had any expenditure to provide house, grounds, servants – anything. But she was satisfied with her accustomed comfort. Ostentation was anathema, and waste a sin.

Morris once complained about the extravagant use of soap in his factories. He had a habit of slitting open envelopes and *using the inside for writing paper. At the height of production he was thought to be earning £2,000 a day. He was created a baron in 1934 and a viscount in 1938, by which time the Nuffield Organisation was producing a quarter of Britain's cars and tens of thousands for export. In May 1939, Morris Motors produced its millionth car.*

Lord Nuffield

And that, shorn of all unessentials, is, for what it is worth, the truth of my early life. I have taken opportunities as they presented themselves, spurred on always by the urge to relieve suffering. It is alarming, almost, how little seems possible of accomplishment, and how much there remains to be done.

The Second World War would wreck his immediate ambitions. His Cowley plant, which had slowly begun to be mechanised in the 1930s, was largely taken over for aircraft repair. His engineers, designers and other staff also joined the war effort, and one of them, who had been with the company since 1936 and was not yet famous, began designing amphibious wheelbarrows for the army.

3 'Somebody hadn't done their planning very well,' said Peter Tothill

Alec Issigonis, BBC radio interview, 1986
My father always had a drawing board in his office, and as a little boy I used to watch him drawing, and when I was quite, quite small I was determined to become an engineer.

[After the war] English people began to emigrate back to Smyrna, and some friends of ours called Smith, the chairman of Standard Oil in Smyrna, they had a Cadillac, and I was absolutely enthralled by this car. They had a Greek chauffeur, and he was my fan because he used to take me for rides in it.

Issigonis was born in 1906 and was British – his father was of Greek descent but had acquired British nationality. His mother was Bavarian. He grew up in Smyrna, on the Aegean coast of Turkey, where his family ran a marine engineering business. A comfortable childhood was interrupted by the war. Then another war, the Graeco-Turkish conflict of 1919–22. At its close, Turkish troops marched back into Smyrna, and the Issigonis family were evacuated to Malta by the Royal Navy. When Alec's father died, he set off with his mother for England, where he suffered from a lack of formal schooling.

I hated mathematics, or anything intellectual like that, because people who are mathematicians are not creative people. I call myself an ironmonger because I do what I do without being hampered by academic atmospheres.

But he found he possessed a talent for drawing. He enrolled in an engineering course at Battersea Polytechnic, where he struggled with exams but found ample diversions. He read the motoring magazines, and joined car clubs and the enthusiast's lecture circuit, where his stooping figure and serious demeanour singled him out. His first car was a Singer Saloon, large and unreliable, £275, in which he and his mother embarked on the Grand Tour. In a manner befitting to drivers enraptured by their early transport, he gave his car a name: Salome. How was Versailles and Monte Carlo and the Swiss Alps? Issigonis recalled principally 'An uninterrupted series of punctures'.

Back in London he became friendly with Edward Gillett, who engaged him in his small engineering firm to work on improvements in gearing that would do away with the need for 'double de-clutching'. In short, Issigonis helped to make driving easier, and more popular. He sold the concept to Chrysler and Rover, and after a few years he came to the attention of Humber, who enrolled him as a junior draughtsman in 1934. The collaborative projects he worked on failed to make it out of the door, but by now he was constructing entire primitive cars as a hobby. With a friend he designed what he called the Lightweight Special, a single-seater racing car he drove in hill-climb tournaments. Issigonis believed that the experience brought two rewards: he learnt the value of regarding car design as a unified solo endeavour, each element joined by the

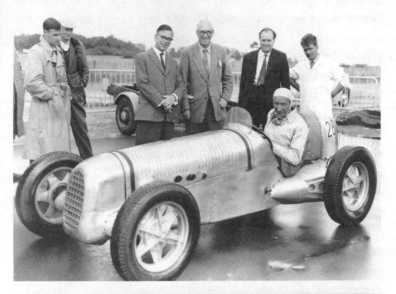

Issigonis in his Lightweight Special and amphibious landing craft

grand scheme, and by necessity by a grand master; and he learnt how to work with his hands. He had very large hands. He was a good sketcher, but his skills were more suited to craftsmanship than the technical blueprint. In later years his friend John Cooper would describe lunches at which their inky tablecloth would disappear at the end of the meal and make its way to the experimental department.

Oddly enough I am a very bad artist apart from design drawing. I can't manage the human form at all . . .

His biographer Gillian Bardsley noted that his sketchbooks contained more than two thousand drawings, and a person is recognisable in only one.

Two years after joining Humber, he joined the drawing department of Morris Motors.

The Morris Minor was reinvented by Issigonis at Cowley shortly after the war, the culmination of his work on the Morris Ten of 1938. Its success was based largely on its ease of use: again, the gearing made it easier to drive than its competitors, and it was dependable and cheap to maintain. It looked like a modern postwar car, its curvaceousness a dramatic leap from the squat and boxy original Minor of two decades before. At one stage, Issigonis decided that his original prototype for the Morris Minor was too narrow, so he instructed his assistants to cut it lengthways down the middle and slowly separate the two halves while he looked on. With four inches between them he told them to stop, and he had his new design. It was launched in 1948.

The Morris Minor on the drawing board and in production in 1948

Issigonis talked of the Morris Minor's 'wheel at each corner' as an innovation – the driver was assured of firm handling and security around bends – but it was hardly an exciting experience: it was an able cruiser, but its rasping maximum speed was 62 mph, and 0–50 took 36.5 seconds.

Lord Nuffield hated the Morris Minor as soon as he saw it, and so did all the salespeople. He described it as The Poached Egg. It was humble pie when we made a million of them.

The Morris Minor was Issigonis's first popular car. Its huge success established his reputation as a frugal, commercial designer who interpreted what the public seemed to want; if you had Issigonis, you also had a very keen eye for simplicity and necessity, and the potential to sell a lot of cars. Issigonis enjoyed the acclaim, and the privileges it brought. His self-belief swelled. He was the only creative figure in the entire British car industry that anyone outside it could recognise by name, and before you could get sixteen students into any of his cars, you first had to remove his enormous ego. His colleagues, judging his swagger unbearable, found a new name for him: Issigonyet. And that was before he drew his plans for the bestselling car in Europe.

Roy Davies
I joined on 3 September 1951, the anniversary of the day war broke out. As an apprentice you did everything. Everybody joined as 'a fitter and turner', and it was a sort of baptism of

fire. They very cleverly adapted one of the shops to make it into a small machine shop. They'd acquired all the equipment from the war, and it improved our French very much, because a couple of the purloined pieces of equipment had been diverted on their way from America to France when France fell. So not only had you got to learn to drive the lathe, you had to learn the French for driving it. You spent twelve months in the apprentice school on those pieces, then another twelve months in the fitting area and then you went out into the factory.

I was born within walking distance of the plant. My father had come up from Wales in 1929 and he had been a driver with a building firm before going to work within the Morris complex. The Morris apprenticeship was the best in the area, and they took on four apprentices three times a year.

I was working on Morris Minors, Morris Sixes, Wolseley 680s, Morris Oxford vans, Morris Minor vans. In those days you had a shop called sub-assembly, where all the suspensions were built, and I can remember the all-pervading smell of the oil that was put into the steering racks, quite a pleasant smell. It was the coldest shop in the factory. There was virtually no heating in it, and most of the doors were open to the constant receipt of supplies, because the suspensions were coming down from the Midlands, and you were handling cold steel, so you froze to death. The one thing it taught you was that producing cars is a very boring process and you've got to learn to cope with the boredom of doing virtually the same thing for eight hours a day.

The one thing that everybody wanted was the belt to break. There was a big conveyor taking these suspensions across, and if it broke you'd be paid waiting time for half an hour. So what you were instructed to do by the old lags . . . A rear axle on a thing like a Morris Minor was a big cumbersome piece on to which the wheels and springs were fitted. It went along on this conveyor, up in the air, and through a hole in the wall that was more or less the shape of what was going through it. If you could induce a wobble as it tried to go through the wall, it would hit and break the shear pin [a fine metal rod], so every-thing came to a stop so we could all rush off to get warm.

Unlike current automation, the bodies were loaded on to the conveyor belts with a measuring stick. It rolled on through the line, and the car was progressively getting its rear suspension, front suspension, engine. With the Morris Minor you had Saloons loaded with a twelve-foot gap and Travellers were loaded with a fifteen-foot gap. But if they'd had a bad run in the morning the supervisor would come down and say, 'Make it a short twelve foot, my boy.' So you'd load it at eleven, and of course being ambitious and wanting to be in the good books of the supervisor you tended to [make the gap shorter and shorter] and suddenly this huge shout would go out from the people up the track: 'We can't get between the cars.'

Peter Tothill

I started in the experimental department in November 1950. I did my engineering apprenticeship at another Morris branch, Osberton Radiators [where they made the grilles for the Bullnose]. Then one of my father's patients, Walter Balding, got me a job at Cowley in the electronics department. A grandiose title for 1950 but there you are. Basically anything that needed

The British Motor Corporation offered 'unprecedented heights' of service from its office in Cowley in 1960

instrumentation, we did. We also got drawn into development work on anything to do with 'noise vibration and harshness' – that was Ford terminology. In those days, the early days of mono-construction, there were a lot of noises and vibrations transmitted into the body shell from the engine, the gearbox, the transmission, the road, and we were trying to eliminate them or suppress them.

Peter Tothill lives with his wife in a large house in Stadhampton, near Chalgrove airfield. He's in his early eighties, and spends time gardening and reading. When he left Cowley in his late fifties he set up his own car parts and accessories business in Abingdon, and since its sale in 2000 he's been heavily

Lunch break at Morris in the early sixties

involved with Abingdon Chamber of Commerce. He walks with deliberation.

That's how I injured my back. We had all the cars in for testing, the Minor, Series II Oxford, the Morris Six, the MG Magnette, the Wolseley 444. I spent a lot of time up at the Motor Industry Research Association [MIRA]. It was formed immediately postwar [at Lindley, near Nuneaton in Warwickshire], and they acquired an old airfield and set about building various test facilities. The pave [cobblestone] track was one of the first, and it was the one thing we used as a means of testing the suspension. We measured shock absorber temperatures because you could work out how much work was going into the suspension.

You did get some quite catastrophic failures. We had a standard that the car had to do a thousand miles without a major structural failure, and I think the first car that did it was the MG Magnette and the Wolseley 444 which had a similar body. At that time steel was allocated on the basis of the number of cars you export. And something like seventy-five per cent of the cars we built at Cowley were exported, not all of them built at Cowley actually, because we had a big CKD [complete knockdown] activity, exporting cars in sets of parts all over the world. And all the thousands of miles of testing we did on those explains why I've got a bad back.

The Nuffield Organisation had established a very thorough objective testing programme, covering performance, fuel economy, noise, vibration and wind noise, structural integrity, suspension and steering characteristics. There wasn't actually a standard goal for anything, we just had to get it up to a particular level. For example, on cooling tests we knew the trouble spots that we'd experience [in export markets] and we'd get an overseas service engineer to bring back some data and then we'd say right, that's what we want to achieve for this particular country. For example, we knew from tests on the Morris Minor that the sheet-metal structure of the front end wasn't strong enough and the shock absorbers, which were an integral part of the suspension in those days, were all right on British roads but quite frankly hopeless overseas. We did a whole series of tests, and the body was modified to take a bigger shock absorber, and then somebody said, 'Well, we haven't had many reports of suspension failures from overseas. You say it's very weak, and we don't disagree with you, but we haven't had a lot of complaints.' So a message went out to all the overseas service engineers, particularly in Africa and Australia, and back came the report, 'You don't think anybody's stupid enough to drive a Morris Minor out of town?'

Morris and Leyland Motors were the main users in those days. Austin used to use it very sporadically and then they were very half-hearted. The Rootes Group, who eventually became Peugeot, used it and various people would arrive, and they thought they were going to do a quick job, and eventually went away without doing the job properly.

Roy Davies
After my apprenticeship I drew the short straw. At the end of your training you filled in a preference list of which department you'd like to go into, and nobody wanted the drawing office because they were always short of people. So that was

always number eight on the list. So I filled out the form like everyone else, and they said to me, 'Welcome to the drawing office.'

If you can imagine you're looking up the layout tables at Cowley, probably two hundred feet in length, and the tables were all in a straight line in the morning. We had a chief draughtsman, and he couldn't bear to look up the tables and see any out of shape. Well of course he was his own worst enemy. So a table would be nudged out of place, he'd look up at it and go, 'Oh.' He would stride all the way up there and try, without appearing to do so, to nudge the table until it was in line and when he got up to the top he'd turn round and look at his office with a satisfied smile and walk back. And of course on his walk back the tables would be shifted again.

When you'd finished at the end of the day you'd say, 'Well that's it then, bye,' and the last person would roll this piece of Rexine [imitation leather] to cover up the tables. We came in one morning and the Rexine was rolled back. Didn't take much notice, thought that someone had been a bit helpful, so we carried on during the day. When we rolled the Rexine out again at the end of the day it only went halfway, because it had been cut off. Jimmy Drewett, the secretary of the Morris Motors rugby team – I was in that, quite successful actually – had cut a piece of it for his own upholstery.

In 1952, Morris Motors Ltd merged with its great rival Austin Motor Co. to form the British Motor Corporation (BMC). Nuffield withdrew from active participation within a year, and the headquarters were established at Longbridge. At the same time, Issigonis was being tempted by the promise of 'a clean sheet of paper' at a rival company.

Alex Moulton

Issigonis had a dominant position at Cowley, but when BMC was formed he could see that essentially it had an Austin lead. He felt nervous about going on at Cowley because he feared he'd be under the subjugation of the technical leadership at Longbridge. So he left to go to Alvis.

They wanted a new car to replace the Grey Lady [a sporty luxurious saloon capable of 100 mph] and they gave Alec a free hand. This is where I come in.

At the beginning of 2009, Alex Moulton, aged eighty-eight, is in the study of The Hall in Bradford on Avon, Wiltshire, with his Moulton bicycle factory up the road.

The Hall is a huge high-beamed house dating from about 1620. Moulton has just returned to it after a spin in the country lanes on one of his small-wheeled bikes with special suspension. His secretary has just arrived too, while his housekeeper has just departed, leaving a lunch of soup and Ryvita sandwiches. He says he is watching his weight. As he sits down in his study he is surrounded by trophies and mementos from his life, including prototype models and bicycle wheels engraved with admiring messages. On the table in front of him is a draft of his autobiography, The Life of an Engineer, *to be published by Rolls-Royce Heritage. There is a postcard from Alec Issigonis describing a trip to the mountains and worrying about the effects of friction on metal. There is also a recent piece on the Mini from* The Times: *'It says Issi sketched it on a napkin – absolute rubbish of course!'*

I had met Issigonis in '48/49. I was fourteen years younger, absolutely dedicated to being an engineer. Issi was already a well-known figure in the motor world, and I was very keen to meet him. The Morris Minor was a great success, and we all had them. They could do sixty miles per hour and go flat out round corners – marvellous. The Frys, those chocolate Frys, were mutual friends, Issigonis had met them through that hill-climbing thing, and I was taken to meet Issigonis at the dining room at Cowley. Driven over by David Fry in his 3.5-litre Bentley. He wasn't at all a teacher or a mentor, but Alec recognised that we had something in common.

I had joined the Spencer Moulton rubber firm, looking for new applications of rubber and constantly designing things. Later I went to meet him at his flat outside Oxford where he lived with his very strong-minded mother, and I showed him drawings I was doing. We became friends, and he had a great sense of humour. But he had no opinion of rubber at all.

But by the time Issigonis joined Alvis, I had established a good relationship with Jack Daniels, head of the research department at Cowley. He was interested in applying rubber suspension to a Morris Minor, so we did one, and we had an idea that it might offer an advantage with regard to road noise, the curse of modern cars. Jack Daniels did the standard thing, putting the car on the test track at MIRA, on the pave, and it did brilliantly. It ran through one thousand miles and the

Normal springs are things of the past

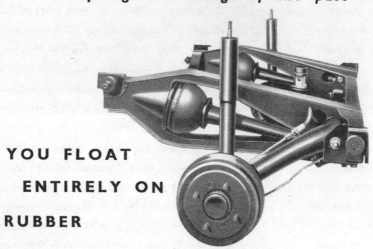

YOU FLOAT
ENTIRELY ON
RUBBER

durability was very impressive, although for road noise it did-n't do any good at all.

Issigonis was both dogmatic and pragmatic, and he observed the evidence, and was then not so supercilious about rubber suspension. He put it on the car he was designing at Alvis, back-to-back rubber cone springs with an ordinary shock absorber in line with it. You could tune the stiffness of it very well, and blend the front with the back.

In 1952-ish I had bought a Citroën 2CV to carry my canoes, and it had mechanically interconnected suspension with coil springs, quite a clumsy thing, but the result was an amazing ride. The boss of Citroën had instructed his people to make a car for a farmer to be able to carry his eggs in a basket across a field without breaking them. There were three of us, the

Three Musketeers, and we used to meet and talk and talk about car design things. John Morris, who became boss of SU Carburettors, part of the BMC group, Alec Issigonis and myself. Our talking led to the birth of Hydrolastic suspension, these rubber cones back-to-back with water in, so we had a fluid interconnection to get the soft ride with good road-holding, and in the Alvis days Issigonis very much accepted the merit of it.

But his car didn't make it. It was expensive, entirely appropriate for the Alvis ethos, but Alvis then decided it didn't have the money. They couldn't afford the new tooling. I remember being at the Turin Motor Show with Alec in 1954/55, and saying to him, 'I'm sure Alvis haven't got the money for your car . . .' So he considered all sorts of simplified tooling, fighting to get the cost down for a mass-consumption car that was actually quite unsuitable for Alvis.

In 1956 he was lured back to BMC and went to Longbridge. Lord realized that Issigonis was *the* man to design the range of cars that he and George Harriman, his deputy, were determined to make. I was one of the people who said to Alec, 'The opportunity is marvellous – do take it.' And he decided so to do.

Lord had been a production and design engineer at Morris, and had risen to become managing director before a fallout with Nuffield hastened a move to Austin. He now resolved to make BMC profitable, and he was the great champion of 'badge engineering', whereby a clutch of disparate models would have an essentially similar body and engine set-up – saving costs but retaining customer brand loyalty. He gave Issigonis a new title,

37

deputy engineering director, and a new office at Longbridge
where he was largely left to his own designs. He was joined by
a colleague from Alvis called Chris Kingham, and Jack Daniels,
his collaborator on suspension systems from Cowley. Those
excluded from this cell, possibly respectful of his talents, prob-
ably envious of his special treatment, called Issigonis the Greek
God. Had he known about it, Issigonis would have agreed with
Balanchine's proclamation that 'God creates, man assembles.'

So we're now in 1956, and he was a bit nervous about commit-
ting himself to Lord. His reputation was great because of the
Morris Minor, but that was quite an orthodox car. The only
innovation was the architecture. The elements were rack-and-
pinion steering from Citroën, torsion bar from Citroën. So he
was now quite glad to have the Hydrolastic suspension. He
explains this to Len Lord, and that brought me on board. The
way I got the contract was that Lord said to Sidney Wheeler, an
Austin man, an accountant, 'Get Moulton up.'

With my brother John we did a big portfolio of all the things
I'd done. So I went up to the Kremlin, and met Sidney Wheeler,
a little man, in the outer office. In we went and Lord looked at
the portfolio and said, 'Hmmm, you've done all this, I'll see you
do some more,' and he turned to Sidney Wheeler and said, 'Fix
him up for exclusive rights.'

We evolved a structure which became Moulton Developments,
which I've still got. In the negotiation I said to BMC, 'You support all my experimental work and pay for it all,' and then in
came Dunlop to do the high-volume production of everything
rubber. Issigonis and I were lured over to Dunlop in Coventry,

and as we were going round the works, Issigonis said to me, 'You are being treated like the Duke of Edinburgh.'

Roy Davies
Issi had created this cell system in Longbridge: some of the people went up from Cowley to be in it and it had typically half a dozen draughtsmen but also fitters who could actually do the things on the prototypes. So there were six people with pencils, and six with hacksaws, and Issi wandering in and out of it. They had to keep themselves alive, so they basically cut and shut what he wanted.

Alec Issigonis
One thing that I learnt the hard way – well, not the hard way, the easy way – when you're designing a new car for production, never, ever copy the opposition.

Alec Issigonis, Alex Moulton and Charlie Griffin solve another tricky problem

Alex Moulton

Leonard Lord said to Issigonis that he wasn't interested in the medium-sized car that Issigonis was proposing, and only interested in a small car. 'It must have four wheels, four cylinders, and you can only use the A-Series 850 cc engine. You get on with it.'

Leonard Lord had expressed a great distaste for bubble cars – 'We must drive them off the streets!' – by which he probably meant the Fiat 600 and VW Beetle as well as the low-powered, three-wheel Reliant Regal and BMW Isetta and the hand-built space-age pods coming out of Italy. He didn't like them as competitors, and he didn't like their design, but he recognised the value of a utility car in straitened times. In 1956, the Suez Crisis had brought petrol rationing. Depending on car size, drivers were limited to between six and ten and a half gallons per month. Perhaps a new British small car would capture the market?

Alex Moulton

It's about that time that those first sketches were done. Not on a napkin – the paper he used was an Arclight [tracing] pad about that size [his hands describe an A3-sized sheet in the air]. He was a beautiful draughtsman.

Issigonis sketched a car that was ten feet in length and used every inch of available space. Passengers were to have as much room as they were used to in a much larger model, and would still be able to take a little luggage. The car would be rigid and safe, but shorn of all luxuries. To achieve this, the engine would be

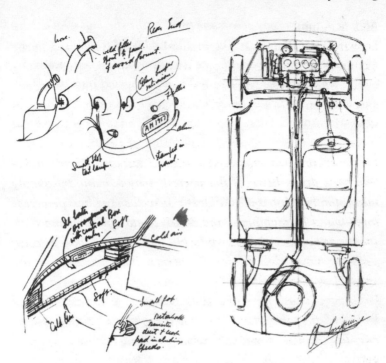

Not on a napkin:
Issigonis sketches
the Mini

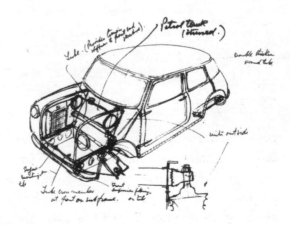

mounted transversely across the front of the car, and would drive the front wheels. The gear change would be placed in the engine sump, and use the same oil. The wheels would be ten inches.

The transverse engine was not new; it existed on some of the cars Issigonis worked on at Humber. The suspension was not new; Moulton had developed it at Alvis. The sump gearbox was unusual, as was the economical design of the interior; it was austerity motoring many years after Britain had seemingly moved into the realm of luxury. But the look and packaging were as appealing as a small metal box has ever been. And the handling was exceptional: the driver, always in control, was in touch with the road the way a Meccano model was in touch with the playground. Given the right tools, it was the car you would have built yourself.

Issigonis was fond of the phrase 'a camel is a horse designed by committee' (he has been credited with its invention, but he may have stolen it), and he wasn't keen to let those below him take credit for their work. It was a pattern of iconic design in this century – the one-cylinder engineer. The Routemaster bus, another great roadworthy icon designed in the decade after the war, is usually credited to a 'design team led by A. A. M. Durrant', but so many other great singular products that we now happily commemorate on British stamps – the Supermarine Spitfire, the K2 red telephone box, the orange-and-white Penguin jacket, the London Underground map – are all credited to one man alone. And the mini-skirt is usually credited to one woman alone, Mary Quant.

There was personal isolation as well. Issigonis lived with his mother until she died, and then alone. He was probably

How it was done: The Mini gives up some secrets

homosexual, he was certainly misogynistic (when told that the fascia on one of his later designs would remind women of domestic appliances, he said, 'I don't want bloody women driving my cars') and he held what his friend Laurence Pomeroy politely called 'strong views on certain foreign countries and nationalities'. He loathed anything that came from America, particularly Ford.

He also held strong views on interior design. Issigonis had never been mistaken for a minimalist before, but his interior lacked most of the features that a prospective customer at the annual Earls Court Motor Show had come to expect. No heater or seatbelts on the basic models, certainly no radio, seats with little padding, everything spare apart from a tyre.

Alec Issigonis

Any kind of driver distraction is to be avoided. It causes accidents. You want to be very alert and sit upright. When I'm driving I don't want people to talk to me.

I never wear a seatbelt. It's much easier to drive without having an accident. Stylists are employed to make things obsolescent. Like clothes, women's clothes. And it now applies to cars. In my cars I made it so they couldn't be obsolescent, to give the customer good value for money.

Alex Moulton

Early on I realised that I couldn't get the Hydrolastic suspension to work in the Mini – the car was too small. So Jack Daniels and myself cooked up a scheme to fall back on the cone spring. We could use a bit of our previous technology, but without the fluid.

Peter Tothill

The suspension system was terribly expensive. In retrospect, if Issigonis hadn't insisted on having rubber suspension, if he'd stuck with conventional suspension, you could have taken pounds out of the cost.

Roy Davies

There was more than one Mini. The Mini we know now wasn't the only car they were thinking of. Leonard Lord had set up this competitive cell arrangement, and I certainly seem to recall on my walking through the block towards the MG drawing office that there were three or four different ones in there. I imagine the brief was to take on Issigonis, be adventurous and do something different.

Roy Davies is at home in a quiet road about five minutes' drive from the Cowley plant. On the table in front of him he has photographs of his time in Sweden as a performance tester in sub-zero conditions, and of Minis that never made it – rounder, longer, less box-like, more fifties-futuristic.

This is one of the 'other Minis'. From memory it had a Goggomobil engine in it and in concept it was rather like a Mercedes racer with gull-wing doors.

The Goggomobil was a tiny Bavarian car with a two-stroke, two-cylinder engine behind the rear wheels, originally 250 cc.

45

A Mini that wasn't: one of Issigonis's competitors breaks cover

It had first appeared in 1955, the design of scooter maker Hans Glas, who later also produced a slightly more powerful small coupé. His company was later taken over by BMW.

It's certainly nowhere near as large a cabin as the Mini, as you can see there. It's not unlike – remember the Nash Metropolitan [an American-designed model produced at Longbridge]? It's a bit like that, and then Citroëny. The only time I ever saw it personally was standing in this private enclosure when it was out on the aerodrome one day.

The picture was taken on Chalgrove airfield, this was the '58 era. And there's the proof that it was roadworthy. 1407DU: that was a plate we borrowed from Coventry when our plates were running out. The person at the wheel is Harry Jakeman, his co-driver is Fred Baker and they were the Charlie Griffin Road Proving testers. At the same time, the airfield was also leased out to Martin-Baker, the ejector-seat people. So the Mini was running round the outside, and the seat ejection was going on in the inside.

Peter Tothill
I saw the alternative quite closely when I wandered into experimental. It was a non-starter really because it wasn't much better than a bubble car.

I was aware of the very first prototype of the Mini being run round Chalgrove, which was only a couple of miles down here, and I'd seen it. Hadn't taken a lot of notice of it, you know, it was secret; you're not supposed to be too inquisitive. And then about that time, Les Ford called me into the office and said, 'Peter, we've got to prepare to build this new car. It's highly secret. It's code-named ADO15 and I've decided I want you to do all the production engineering.'

I had enjoyed my five years working in experimental, but I decided it was a bit of a backwater and there was a vacancy in production engineering so I thought I'd go for that. And got it. So I was given responsibility to look after the Wolseley line – there were four assembly lines in those days – and I discovered I'd taken over a horrendous mess. The bloke I'd taken over from was just bloody idle. So I had a lot of major assembly problems, and I sorted all those out – well, ninety per cent of them.

The new car, ADO15, was of course the Mini. They said, 'The first thing you'd better do is get up to Longbridge and go and see Jack Daniels. You'll know Jack, because he used to be at Cowley.' So I went up and Jack showed me the car, showed me where they'd got to, and then when I came back I was shown the area where we were going to build it. That would have been the back end of '57.

A few years previously we'd built a new paintshop at Cowley, and the intention was to have two paint lines and two rotor dips. A rotor dip was the rust-proofing process that was carried out in those days where they dipped the whole body into various vats and the body was mounted on a spit so it could be rotated. But somebody hadn't done their forward planning very well, because they decided that we didn't need a second rotor dip so we'll send it to the Australian plant. So there was this big area – not big enough as it happens – of bare concrete floor. I looked at it and said, 'This isn't half the length of the existing assembly line, how are we going to do it?' My

An orange Mini prototype with a deceptive Austin grille

boss Les Ford said, 'That's your job, but it's going to be a very simple car.'

Issigonis wouldn't let you alter anything. I'd been getting component drawings coming through, and I was collecting a pile of those which were of little use to me because there wasn't a full-sized layout. So I spoke to my boss and said, 'This is hopeless, I can't check anything, we need a full-sized layout.'

'There isn't one at Longbridge. Issigonis's team haven't done one.'

There was quite a big drawing office in Cowley, so a couple of guys, George Cooper and Ron Unsworth, were given the task of doing a full-sized layout. They would do a side elevation as well, and bearing in mind this all had to be drawn with pencils – there were no computers in those days – it was a long process. Sometimes the boards were vertical and you stood up at them, but this board was horizontal and they were both pipe-smokers and the whole layout got black with ash and burns and sometimes it was a bit difficult to read, and it was made worse by Issigonis. He used to come down from Longbridge on a Friday because his mother lived in Oxford still, and he would come over to George and Ron and they would discuss things with him. He always used a very soft pencil and he'd get his pencil out and start scrawling all over the layout. Once they cottoned on to this they put a piece of tracing paper over anything they wanted to preserve.

Roy Davies
George Cooper was a very fiery character. Very nice chap but a spade was a spade with him. Of course, when we went to put it

together and it was 'Sorry George, it's not fitting,' he'd tear his hair out. Issi's attitude to life was that nothing is wrong. Anything that doesn't fit is a challenge to you. But you can't have a challenge if your drawing doesn't go together, can you? And so there were one or two spats between the people from the cell, and Issi would come in to referee it.

George was the section leader, and Ron Unsworth was his assistant. It was the tall and the short, or the thin and the fat. Ron was an ex-RAF pilot who used to fly Liberators, and they had plenty of room inside them. And George just used to come out with these comments, whether it was Issigonis he was talking to or whomever, he'd say, 'This will not bloody well fit the engine! The engine's up through the bonnet!'

I was fortunate to be allocated to his section as a trainee detailed draughtsmen. These drawings kept on coming down from Longbridge and it was 'Detail this.' Well, where shall we start? There was no reference to where it actually fitted relative to the car. We then laid it on the big sheet and found that it was sticking out through the wheel. The people at Longbridge would have modified the wheel bracket, but had forgotten to tell you.

So this is where the rancour occurred between Longbridge and Cowley. There was very little interaction between Longbridge engineering and Cowley engineering. Longbridge were solving their problems and then telling us, 'We haven't got a problem,' and it wasn't until somebody physically went up, Peter Tothill or somebody like that, and said, 'Well, the reason you haven't got a problem is because you don't use this bit any more . . .'

'Oh no, we got rid of it.'

It was a peculiar situation. It was the first model that we ever built at Cowley that wasn't designed at Cowley, whereas before you could always pop into the drawing office and say, 'Why's this?'

Peter Tothill

The prototype I saw wasn't greatly different from the car we know now. None of the fundamentals changed. It was ten feet long, it was the same width, it was low, it had ten-inch wheels, the suspension was basically the same. But they did have to turn the engine through 180 degrees. There were two reasons for this. The drive to the gearbox, which was in the sump, had one very large gear on it and it was noisy, and for some reason they wouldn't use a chain. I don't know why because motorcycles always use a chain. And the other reason was, they were getting a lot of problems with icing.

Roy Davies

In about March 1959 [five months before launch], some of the Cowley testers suddenly began reporting that they were getting peculiar failures on a morning like this – when there was a frost about. The car would suddenly stop, or the accelerator pedal would stick, or they couldn't get the car to start at all. But by the time they lifted the bonnet it was all perfectly OK.

So one of the first things I did in the proving department was finding out how many petrol companies there were in the country, Esso, Shell, Regent, and I think it came to about ten. So we had to get ten new production Minis, put ten new petrol tanks in them and brand-new pipes, and each car had to be run every night – an Esso car, a Regent car. And the failures were completely random, they failed on Esso, they failed on Regent, and some didn't fail at all. And all it came down to was this icing-up problem. The reason for it was because a late decision was made to turn the engine round from its original position, thinking that putting the carburettor at the back would protect it. But it did the very reverse, it froze it, because it was sitting in a pocket with no air coming over it. Nobody had sort of thought through, 'I wonder what's going to happen to the carburettor.' Nobody ever told Issigonis, because he didn't listen to anybody. The problem was finally solved by ducting warm air to it from the exhaust area.

There was an environmental ice chamber in the experimental department at Cowley where start-ups were done regularly on the normal cars. But the Mini didn't come down from Longbridge to have that done to it.

Peter Tothill

Once the prototype had gone back up to Longbridge I set about producing an assembly precedence list. You know, we've got to fit this bit before we fit that bit.

I went to the drawing board and laid out the space I'd got and started drawing little silhouettes of the car and decided, well, the car's ten foot long, we've got to have a four-foot gap between that and the next car, and I worked out how many cars I could get in the space and thought, ouch, that's not enough.

So I decided we'd have a little preparation line for the sub-frames and then transfer them on to the assembly line. My next task was to draw matchstick men on to simulate where a person was working – above, below, at the back or the inside – so you didn't have everybody crowding around a car that they couldn't get at. We reckoned that the facility would do twenty an hour. In fact I don't believe it ever got above fifteen an hour. It began on a single shift, and then moved to the double shift. Sixteen hours, five days a week. I think our target was sixteen hundred cars a week.

The one thing I always did very carefully . . . You would have a chassis section who did all the mechanical components; then you had a body section who'd do all the trim and finish. The two had to come together and I was always very careful to check the clearances. A bit of the conveyor had a break in it where the body would come overhead and into the drop section. You only had about six feet, ten at the outside, a very limited time with the track running to get it down and bolted. And that was the first major confrontation that we had with Issigonis.

When we had a second prototype we discovered we couldn't do a straight body drop, and I reported this to my boss and said, 'We've got a major problem.' And so a meeting was arranged with Issigonis on a Friday afternoon. There was Les Ford, Harold Cross, who was the Longbridge production engineering man, myself, who gave a demonstration, Issigonis and Jack Daniels. So I gave the demonstration, and Issigonis took the attitude, 'I've designed it, it's your problem to put it together but you can't alter anything.' We said, 'We can't do this, because as you lower it, the steering rack, which was bolted to the toe board on the bottom, fouls the back of the cross-member of the front subframe as it passes.'

'Oh, you'll have to jiggle it.'

'We can't jiggle it. It's got to come down – bang.'

Issigonis adopted his usual attitude, so Harold Cross said to him, 'Well, I'm sorry you're adopting this attitude, Mr Issigonis, because you're leaving Les and me with no alternative to report back to our respective directors on Monday morning that this car can't be built at Longbridge or Cowley.'

So there was two minutes' silence only broken by Jack Daniels. He was a lovely guy, I had a huge respect for him. Jack was a pipe-smoker and his thinking time was very simply – he'd get his pipe out, stir it up and light it, and then he'd reply to whatever you'd said. So anyway, Issigonis turned to Jack Daniels, said something that no one else heard, and stalked off. And Jack said, 'He's told me to do what you want.' So we discussed how to overcome the problem by redesigning the rear crossmember of the front subframe. That was Issigonis. He didn't want to be involved in it – he left Jack to do it. It was obviously a serious loss of face for him and after that he would barely talk to me.

I was in Jack's little office one day and Issi comes in, looks at me over the top of his glasses: 'What are you doing here?'

'Just come to discuss a few problems with Jack, Mr Issigonis.'

'They're not our problems, they're yours. Go away.'

Whenever I went to Longbridge they treated me like 'one of those blokes from Cowley with straw in their hair'.

On subsequent times when I went to see Jack I always used to walk into the Longbridge assembly line and look at the points I wanted to raise with him. And during that time I got to know Dick Perry. He was the general superintendent of the assembly building in those days – he eventually became MD. And then he left and was MD of Rolls-Royce. I bumped into him a few times and he said, 'What the hell are you doing here?'

So I explained and I said, 'Have you got a problem with so-and-so?'

'Oh yeah,' he said, 'I've solved that.'

So I asked, 'Who knows about it?'

And he said, 'Oh, I've just done it. I don't worry about all these other people.' So we had another autocrat building the thing, and if something wouldn't fit he would just have it altered. He'd get on to the supplier and say, 'Stop making it like this, start making it like that.'

And so we had the suppliers working on two different conditions, and the Minis from Cowley and the Minis from Longbridge were detailed differently. One example was, we had terrible trouble fitting the windscreen because it was intended as a dry system, you didn't have to use any sealant. You had a rubber that you fitted to the glass and then you put the glass and the rubber into the body aperture and got the flange of the rubber over the flange of the metal and then from the front, with a special tool, you inserted a sealing strip that spread the rubber to create a tight joint. And we'd not got a good rubber. It was inflexible and we were having terrible trouble with this. Cars were going off with the filler strip not fitted and all that sort of thing. So I said to Dick, 'What have you done, your windscreen goes in, I've been watching them.'

He said, 'Oh, I got on to a local supplier and got them to change the specification of the rubber and they fit fine now.' Hadn't told anybody.

I went cap in hand to the buying department at Cowley, and they'd come back and say it costs another sixpence. I said, 'If you want to know what our rectification costs are I'll find out, you know . . .' So eventually we got that changed.

Roy Davies
Then there was the painting. When the Mini first came to Cowley, there was a man called Dick Couch, the works director I think at the time, and he said to the paintshop man, a man called Dick Price, 'Well Dick, I have to tell you this. You're going to paint the Mini for a fiver, and' – because as the body shells went out of the paintshop, they passed an office which the supervisor sat in – 'you'd better make sure that you take a big wide brush out of the window and as it passes, paint the supervisor's window. Because we're not spending a lot on the painting.'

As it turned out, it was an absolute nightmare to paint because the roof was cleverly designed to shed the water off it, so the water on the roof came into the gutters and dribbled down the back. But so did all the paint. When you were painting the roof, the paint used to start coming down out of these little bird beaks, as we'd call them. And the Mini had these famous flanges on the side, and they were dreadful because two things used to happen with those. They would constantly upset your paint flow and you couldn't get paint right underneath. So of course you've got surface rusting in there. Even things like Mini wheels: we had this thing called electrostatic painting where you effectively threw paint into the air, put a charge into the wheel, and the paint came down. Well of course the Mini wheel was so small you couldn't throw the paint into it very easily and that took a long time to resolve. But lots of the Mini problems were because it was different, not because it was difficult.

Peter Tothill

During the summer of 1959, we'd built twenty or thirty cars maybe, obviously it was very slow to start with. And they're all lined up and there was one afternoon when I was over there, because I practically lived on the assembly line for the first six months, and there was a thunderstorm. I thought, 'Ah, leaks.' Because the water came in off the road, not from above. So I found the tester and said, 'Get a set of plates on a car, we're going out.'

'What, in this?'

'Yes.' So anyway, we got a set of plates, dashed out, got into the car. He said, 'Where are we going?'

I said, 'Up to the Headington roundabout and back.'

He said, 'Why, what are we doing?'

After we'd done less than a mile he said, 'My feet are getting wet.' By the time we got back, I suppose it's probably about three-quarters of a mile, we'd got two inches of water in the car. There were no carpets in at that stage because they were always put in right at the last stage, just before the car was dispatched. And so he said, 'Well, I didn't know they were going to leak like

Opposite: The splash test – life jackets optional

that,' and I said, 'I did. It's going to run out as fast as it's going to run in, so what we need to do is to get this car to the photographic department,' which was just over at the next block. And so I whizzed in, got hold of one of the photographers, 'Are you busy for a minute? We want our picture taken straight away.'

He said, 'What is it?'

I said, 'We've got a Mini outside and we've got two inches of water in it.'

So he comes out and has a look and says, 'If I take a picture of that it won't show.'

So I said, 'Oh, what are we going to do about that then?'

Anyway, we go back in and find a bottle of ink, so we go back out and tip the bottle of ink in, mixed it all up and we'd got photographic evidence then.

Then there was panic. Big panic. So the first thing we did was build a water splash. I said, 'What are we building a water splash for? We know the bloody things leak.'

'Oh well, they may not all leak.'

So I wonder what happened to the photographs?

God knows. God knows.

Laurence Pomeroy (*The Mini Story*, 1964, written with Issigonis's cooperation)

The earlier examples of ADO15 . . . were notorious for water seepage into the floor with consequent soaking of carpets and emission of a strong musty odour. The defect was as unfortunate as it was unexpected, in that the passenger compartment for the prototype cars had been as tight as the proverbial drum, and indeed the cars had been driven repeatedly through quite deep water splashes without any trace of a leak . . .

The fact that this trouble was not discovered before the production period was largely occasioned by the exceptional summer of 1959, during which the production prototypes were driven week after week on dry roads. It was a further misfortune that the winter of 1959–60 [after the car's launch] was exceptionally wet.

Roy Davies

A lot of the problems were self-generated by a refusal to accept that there was a problem. I mean you didn't really need to go to Sweden to know that the heater wasn't working properly. It just wasn't man enough for the job. And you knew that the

Taking the temperature in sub-zero Sweden

Testing the Mini, Rover and Triumph in the snow. Roy Davies is in the centre in a tie.

A Mini prototype on test at Chalgrove airfield. Pat Cox from Cowley
Experimental department in attendance.

demisting was inadequate. But remember, even when you're in
the road-proving area, and you're doing, say, endurance testing,
all you're trying to do is get the maximum number of miles out
of the car, so you're not really interested [in other issues]. So you
might say it's cold, but it's not a priority that it's cold. You just
took a coat and did the old-fashioned trick of a potato on the
glass to stop it freezing up, and you got over it that way.

With the water leaks, the people making the body knew that
it was being built upside down. It would have made wonderful
television. Poor old Bob Lambert, the chief body experimental
engineer. When he went to the Christmas do, the chef came out
with a large plate full of water, and all the way around the edge
were all these little Dinky Toy Minis. He put it in front of Bob
Lambert and said, 'Now you watch this, Bob, it's just like the

real Mini,' and he tipped the wheels, and the cars rolled gently down into the water and sank.

Peter Tothill
They decided to fill the sills with foam and they used a product which was highly toxic and it had to be stopped for health and safety reasons. Although health and safety hadn't been invented then.

Roy Davies
The brakes were a problem. Not because anybody had badly designed them, but on a front-wheel-drive car, if you've got any

imbalance, if those brakes are not quite bedded in, it puts twice as much into one wheel as the other. Once the brakes were bedded they weren't too bad at all but in the early days they would dart and weave all over the road.

The other problem which caused us to drive round and round Chalgrove airfield was that the noise from the fan was like a banshee. You didn't need a horn; you could hear it coming from miles away. The problem was keeping the cooling performance going, because you could chop a lot of blades off the fan and it would be wonderfully quiet, but then it would boil like a kettle. It didn't really get resolved until the advent of the electric fan many years later.

The engine had been put that way to suck air through the radiator, so the fan used to scream. But that was another dismissive thing from Issi: 'Don't worry about the noise.'

A lot of gearbox problems with them. I worked for one of the quality directors at the time, a man called Eric Lord, and he was driving backwards and forwards to Innocenti [the Italian motor works which made its name with the Lambretta scooter], who were progressively taking over the manufacture of the gearbox. I remember Eric Lord once came back with his tail between his legs, because, he said, 'I had a very embarrassing meeting over there.' What was that then, Eric? So he said, 'Look, here's the chart . . .' The closer you can get your gears to the norm the less trouble you've got. He said, 'This is what they showed me, look. This is the spread of gears, this is the centre line. These are the acceptable, these are the rejects. These are the ones being made by Innocenti, up to there; these are the Longbridge gears . . .' He said Innocenti wouldn't accept any of the Longbridge gears, even as rejects, they were so bad on the machining on the gear teeth and things like that.

And in the very early days of proving, the clutch was in an awful lot of trouble, again not as a clutch, it was a quite brilliantly designed clutch, but the oil leaks, because the engine would leak on to the clutch plate and then you were in all sorts of trouble.

In fairness, things like oil leaks on to clutches really developed once you'd launched it. They wanted to get it out as soon as possible, and some of the standards that applied to other cars didn't apply to the Mini. One of the Wolseleys, for example. One day Charlie Griffin, the head of road proving, came in to see his boss, Walter Balding. Now Walter Balding was a very strange gentleman. Quite short, very gentlemanly, straight, and nothing would move him. Charlie Griffin said to him, 'I could get this car to go round a corner without turning the steering wheel.' Walter Balding said, 'You'd better demonstrate it to me, Charles.' Charlie went hurtling down the road, which goes out towards Stadhampton, and there's a turn-off. He judged it nicely, hit the brakes and went straight up the turning. Walter Balding brought it back, had a look at it, and it was just that the front end was too weak. So Walter said to Charles, 'Put it in the corner, we'll sheet it up.' He said, 'We're stopping all the work,' and said to the designer, Gerry Palmer, 'It's not coming out from the sheet until you make some modifications.' And so they made the modifications and away we went with it. Now the Mini certainly didn't go through that, because the cell system was so bitty, because you really weren't quite sure what was being designed at the other end and what was coming through.

Peter Tothill

In retrospect I think it was a bit too far ahead of its time. It was rushed through on the basis of fuel economy, although in fact it wasn't that much more economical than a Morris Minor.

I remember very clearly indeed Les Ford, my boss, came to me on the Wednesday before Easter and said, 'Peter, we've got enough bits collected up to build the first pre-production car and we've decided we want to build it over the Easter weekend because there'll be nobody in the factory. It will still be completely secret and so I want you in all over the Easter holiday.'

I hadn't been married that long and I had to come back on the Wednesday and say to my wife, 'Our plan to go away, it's just been hit on the head.' And she remembers it too. And so that white Mini, the first Mini [621 AOK] was definitely built at Cowley.

It's a Morris and there was a very distinct division between Longbridge and Cowley in those days. Morrises were built at Cowley, Austins were built at Longbridge. That was broken because Morris never built the Traveller; they never built the Mini Cooper. They were always built at Longbridge. I can assure you that that white Mini was built at Cowley. Longbridge claimed it, but eventually I believe they've acknowledged that it wasn't built at Longbridge, it was built at Cowley.

It was transported on a lorry up to Longbridge on the Tuesday, put into the design studio or wherever it was for the directors to see on Wednesday morning. That's why it has a Birmingham registration. We'd got it all done by the Monday, but we'd got one or two touch-up paint jobs that needed doing, and so we took it over to the experimental paintshop and muggins here brushed his jacket against the wet paint, which didn't go down very well. The guy had to redo it and I'd got paint on my jacket.

And that was the first car?

Yes, that was the first car.

A bit too far ahead of its time: 621 AOK, the first Mini

4 'In that case, I'll have a double whisky,' said the man from the press.

Roy Davies

The press launch took place at the British Army Vehicle Testing Establishment at Chobham in Surrey. You're used to hearing that it cost £10 million to launch this or that, but I think it was £2,000 or £3,000 to launch the Mini. We got all these cars ready and from memory the only colours were red, white and blue. The normal thing was you ran your cars in for five hundred miles, sorted all the bugs out, then we went down to Chobham, the place was all sorted out, ready to go, and the press gathered round and it was like looking at large go-karts whizzing round. They'd never actually come across a car like it. It was a flat test circuit, there was no banking really on it to give you any benefit, I mean they were whizzing them round there virtually flat out and they just liked it. You could do something like 70 mph.

I was in the press tent. You had to hover in case anyone had any questions. This chap walked up to the bar and said to the barman, 'I'll have a half.'

The chap behind him said, 'Don't you realise you're not paying for this stuff, old boy?'

He said, 'Well in that case I'll have a double whisky.'

Letter from a French motoring correspondent

I wanted to tell you that I enjoyed your hospitality and fine

WORLD'S MOST EXCITING CAR

with engine mounted

FAR
MORE
ROOM
IN FAR
LESS
SPACE

EAST WEST

From the first Mini brochure

organisation. The whole BMC presentation was beautifully organised, and I did appreciate the many details, without forgetting the admirable Chateau Margaux which made the whole thing a memorable occasion . . . Mr Issigonis is to be congratulated again for all the technical features which beat all of the other cars in its class . . .

A few days after the motoring press drove round Surrey, a more spectacular event was held for dealers and the general press at Longbridge. The publicity office had a £500 budget, and what didn't go on hospitality went on a giant top hat. The theme was wizardry: how was it possible that eighty per cent of a new car was habitable? The hat was lined with silk, and as the dry ice and music swirled around it, it tilted upwards to reveal a Mini. A publicity man familiar with the ways of P. T. Barnum waved a wand.

Tony Ball
Inside the car I had put three of the biggest men I could find, two ladies, one of whom was my wife, a baby, who was my three-month-old son, golf clubs, two rather large poodles and all the luggage we could possibly cram in, in door wells and under seats . . . People just stood and cheered.

Journalists left Birmingham with a copy of the Mini's first sales leaflet, or the first two: Cowley and Longbridge had their own marketing departments, and each produced their own literature and promotional films. The Mini only became 'the Mini' after years of common usage by drivers and the press (and only after headline writers stopped calling it 'the Baby'). It was initially

Poodles – check. Golf clubs – check. Three large men . . . The Mini
meets the press, late summer 1959

*called the Morris Mini-Minor and the Austin Seven (or, in many
proud advertisements, the Austin Se7en).*

*Essentially these were the same car: the radiator grille and
colour shades differed slightly, but which car you bought largely depended on brand loyalty: you were either a Morris driver
or an Austin one. Build quality was just as variable whether
made at Cowley or Longbridge. The promotional literature and
guide books each pretended the other brand failed to exist.*

*Both of the initial marketing campaigns shared one theme:
size. The size of the exclamation mark.*

The first Morris sales leaflet

The car that belongs to the future – belongs to you!

Built to give you big car motoring. Big in Performance! Big in Comfort! Big in everything but cost and upkeep.

This is a car which began with an idea. An idea that would double the pleasures of motoring for millions!

Never before has such quality, performance, economy and sheer reliability been offered to so many – at such a low price!

The Morris Mini-Minor is the most downright *sensible* motor car ever made.

It carries four big adults and as much luggage as they're ever likely to need. It has a full-size 4-cylinder engine of proven design which leaves you with 10 mph in hand even when cruising at 60! Its cruising fuel economy of beyond 50 mpg means pounds in your pocket. [When the car was launched, petrol shortages were no longer a problem.]

It's easy to see how much space is saved by the brilliant 'East–West' engine. But that's only the first touch of genius.

Gearbox, transmission, steering, suspension are all revolutionised; simplified and improved to a degree that the motoring world barely dreamed possible.

The Morris Mini-Minor is the first fundamental change in motor-car design for years. It is the start of a new era. Here is tomorrow's car waiting for you today!

The brochure, which folded out into a poster, was called 'Wizardry on Wheels: The Revolutionary Morris Mini-Minor'. 'Wizardry' was placed on a flash of jagged sunburst colour, like the Daz symbol. The era in the accompanying illustrations was

THE
REVOLUTIONARY
"QUALITY FIRST"
MORRIS
850

mythical, a time of picnics and cocktail parties, of hair-bobbed children in endless bliss. Suez had been forgotten, the land of plenty and security returned. The blue car had two smiling people in the front and three large ones in the back and the text that followed gave many reasons why they were happy. The first was, it wasn't raining.

Space-making compactness of design GIVES parking as easy as winking. In busy streets the Mini-Minor simply dances through traffic – and it nips in and out of parking spaces where most drivers wouldn't even try. This superb manoeuverability comes from the fact that it's only 10 ft in overall length – the roomiest car ever built with such compact measurements.

Four-Wheel independent rubber suspension – new springs – GIVES long-distance comfort and simply stacks more room. You'll rub your eyes with disbelief when you see how much luggage she carries; and you'll clap your hands with joy at the beautifully effortless ride this amazing motor-car gives you.

Front-Wheel drive does away with ordinary transmission – GIVES effortless motion – like a ballerina! In many motor-cars, a large transmission shaft runs from front to rear and there is a big differential 'hump' in the back axle. For the Morris Mini-Minor, genius plus common sense has produced an entirely revolutionary system, *independent on all four wheels!*

Servicing is three times easy! [Picture of oil can.] The same oil is used for engine, gearbox and differential.

The simplest, most effective windows! They simply slide easily to and fro for draught-free fresh air. Look at the expanse of rear window [picture of woman in red jacket with various things on the parcel shelf behind her, including a book and a camera]. She's sitting pretty – with a perfect view all round!

The front seat tips lightly forward to give the easiest possible rear access. No crumpled frocks with roominess like this!

Tailored inside like a magician's top hat!

Remember the times – on holidays and picnics – when you've longed for twice as much room. Well now you can have it!

You'll bless wide doors at times like this! [Picture of a man in evening dress sitting down, with a woman in a necklace and white gloves outside the car.] The Morris Mini-Minor gets around in society; wide-opening doors and extra leg room ensure elegant exits and entrances on special occasions.

Fast runs for opening batsmen! [Picture of man in cravat

The Mini goes a-marketing in Tudorbethan England

SPACE-PLANNED FOR CAPACITY

Deep door pockets. Here's where you put maps, handbags, flasks, orange squash, and all the dozens of odds and ends you need on long journeys. And they're handy on shopping trips too—so easy to fill and so quick to empty.

The luggage trunk? Actually inside the car! Simply open the flush-fitting trunk door and you'll find room for everything you need carry.

Why carry it on your lap? Travel in comfort. Put your personal belongings under the rear seat along with the picnic hamper and other odds and ends.

reaching into the car to place or retrieve a leather cricket bag.] You could carry enough gear for a test match! But even though Dad may only be a spectator, he'll get to the game in double-quick time at 50 mpg!

Engine: 4-cylinder, bore 63 mm, stroke 68.26 mm, 848 cc, developing 37 bhp at 5,500 rpm, compression ratio 8.3:1.

The early reviews were flattering. 'The most sensational car ever made here,' proclaimed Courtenay Edwards in the Daily

Mail. '*After testing it yesterday I can say confidently that this car . . . will start a motoring revolution.*'

'*Right for the family and any kind of traffic,*' opined Basil Cardew in the Daily Express. '*I reached 60 mph in under half a minute. Clearly the car will outstrip many on the road.*'

The size was news, and most of the reports mentioned the transversely mounted engine, front-wheel drive and independent suspension. '*It's a people's car that opens up a new era in family motoring,*' reasoned Tom Wisdom in the Daily Herald. '*For two days I hurled the little car through every imaginable test, ending up with a real bashing on the Army's testing ground at Chobham. I gave up. I couldn't fault the machine.*'

'*A new era in democratic motoring,*' suggested W. E. A. McKenzie in the Daily Telegraph. '*The BMC twins – a new Austin Seven and Morris Mini-Minor – offer the artisan, clerk, and wage-earners at large a "people's car". They provide motoring without compromise at a penny a mile.*'

'*The suspension is quite remarkable,*' calculated the motoring correspondent of the Financial Times. '*I drove the car over deep potholes in a broken-up concrete road and hardly felt a jar.*'

'*Wizardry on wheels!*' agreed Charles Fothergill in the Daily Sketch. '*That's what I think of the BMC's long-awaited baby cars. I found road-holding magnificent.*'

A more sober appraisal came from The Motor, *whose correspondents had been given an early car to drive from the Midlands to Somerset and Devon. The drivers – described as 'two far from light men' – managed a maximum speed of 72.4 mph, 0–50 in seventeen seconds, and a touring fuel consumption of 43.5 mpg. Sometimes the drivers took on two more adults. 'The miracle of passenger-packing' was credited to the clever engine design and the small wheels at the extremities eliminating intrusive humps.*

The Motor, 26 August 1959

Getting into or out of the rear seat, past front seats which tilt bodily forwards and 'stay put' until pulled back into place, is less dignified than entry to higher built cars. Once a passenger has stooped to enter, however, he or she finds the back seat astonishingly roomy and comfortable. Praise is due to a backrest subtly shaped to steady two people during cornering . . .

Whichever seat it is judged from, the new Austin Seven is a car which provides an excellent outlook, through big windows framed by slender pillars. It also has immense amounts of stowage space for picnic provisions . . . The unconventional all independent springing of this car by means of rubber cones produces a somewhat 'continental' effect, the springing being fairly firm, but immune from 'bottoming' and with low unsprung weight to eliminate shock on rough roads . . .

Some people are at first rather shy of the unwontedly light and responsive steering, but in fact this is an immensely controllable little car and physically quite untiring to drive for long distances . . . It would be hard indeed to spin this car by cornering too fast . . . Fast drivers will find the handling of this car very much to their taste.

Various types of petrol were tried, from 82 to 90 octane. The lowest grade showed a certain amount of 'pinking' when wide-throttle openings were used below 40 mph, but this in no way necessitated the use of 98 premium grade fuel. Even

Everything spare apart from a tyre: the interior's simplicity of line and purpose

laden, the car managed tricky starts sweetly on Porlock and Lynton hills.

Transmission noise is not altogether absent from this car, but is evident mainly at the extremes of the speed range . . . The gear-change by a rather vague swept-back central lever, is far from being the best feature of this new model . . .

Publishing an early test report upon a brand new model, it is sometimes necessary to make reservations about curable detail faults. Our first intensive trials of the new Austin Seven have shown singularly few such imperfections and make it obvious that this compact car priced at only £537 6s 8d offers a remarkable combination of speed with economy, roominess with compactness and controllability with comfort.

British Pathé News, 27 August 1959

Motor News from Surrey! [Usual twinkly patriotic music accompanies clips of men in shirtsleeves looking at new cars.] Newsreels and press men went to Chobham to see the new 850 cc Austin and Morris cars. Designer Issigonis has given them an attractive look. [Issigonis, in buttoned suit, carries a walking stick as he peers beneath the bonnet.] The engine goes across the frame to provide front-wheel drive. Not even the heaviest passenger can daunt it. [Fat man in high-waisted trousers heaves himself into the car and drives away.] These two new British Motor Corporation models have two aims: to keep out foreign cars in the small field, and to carry the sales war into foreign countries. On the rigorous test circuit they showed their hill-climbing powers: the gradient was one in four. [Two cars charge up a hill and then tear round corners.] The little cars corner at

Not for the first time, the Mini takes on the world

amazing speed. They look like showing the Volkswagen and similar invaders just where they get off.

John Cooper

People laughed at the Mini. When he first produced it, they thought it was a bit of a joke, they thought this man's a bit of a comedian. It wasn't until you drove it, and realised how much room there was in the car and how it handled, that you realised this was a serious motor car.

Alex Moulton

At the launch, Issigonis was very impressed by how the foreign journalists were impressed. In-house, the Austin people were all 'that funny little car . . .' and weren't encouraging him. He was very much flattered by the attention. His mother was very proud of him, and he was beholden to her.

83

At the 1959 Earls Court Motor Show in October, the Morris Motors stand was situated between Peugeot and MG. There were four Minis on show, with variations in colour and upholstery and a left-hand-drive export version. Next to them, the cars being overtaken: a light green Morris Minor 1000, a Morris Minor Traveller in blue and grey, a Morris Oxford Series V in maroon and red, and a Morris Oxford Traveller Series IV in blue and off-white. The Minis had their own colour scheme, and seen from the gantry above they represented a symbol of patriotism that was to mark them for their entire life: they were red, white and blue. (The cars produced at Longbridge had slightly different colours, and a more defined purpose: 'Tartan Red, a rich distinguished colour; Speedwell Blue, a pastel shade very popular with the ladies; and Farina Grey, well known for its hard wearing qualities.')

The cost of this engineering was:

Standard: £350 plus £146 19s 2d purchase tax.
Deluxe: including a heater and over-riders, grey fleck and black upholstery, and dark grey, cherry-red or blue carpet to suit the body colour: £378 10s plus £159 16s 8d purchase tax.
The optional heater on the standard model cost £6 10s. The optional extra radio cost £18.

The problem was, it was all too cheap. The story goes that people from Ford's planning department bought a standard Mini, *took it apart, and tried to work out how Morris or Austin could make any profit from it. Then they realised they couldn't. It was selling at below cost price, and BMC was losing about £30 on every car. Surely things would change within a few weeks. But they didn't.*

The model had been priced before the prototype had been made, and thus took no account of the high-performance engineering Issigonis demanded. The car was spartan, but the suspension engine configuration and marketing budget was not. The Mini was launched on the assumption that it would only succeed if it was the lowest-priced car in its market. But the Mini was not merely a designated loss-leader. An investigation by the management accountants Cooper Brothers revealed that BMC – having long absented itself from the fierce business acumen of William Morris and Herbert Austin – was lacking the fundamental ability accurately to measure the cost of any of its cars.

Those who attended the Earls Court show were offered a new promotional booklet. On the front, four young adults were enjoying a picnic next to their red car, and their life had improved beyond measure ('It's Exciting! The Car of the Future – Now!' etc). But inside the booklet the tone had changed. It had a more corporate air, and a faintly defensive one. Its writers were no doubt aware that Ford also had a gleaming model at the Motor Show, the new Anglia.

The second colour booklet

The British Motor Corporation offers the most convincing vindication so far of its policy of component production rationalization and parts standardization.

Its announcement fulfils a prediction made less than two years ago at a Motor Show luncheon in London, when Mr George Harriman, managing director of the British Motor Corporation, stated that the Corporation's statisticians and market research experts had formed the view that the public did not want bubble cars but a low-priced fully-engineered car of excellent performance. 'Obviously,' he went on to say, 'if the Corporation can produce such a car which will sell more cheaply, they will do so.'

Here, then, *is* that car. It positively sprouts with new ideas like berries on a holly bush.

There was a familiar list – the transverse engine, the suspension, the downward bootlid, 'the deep pen glove boxes for all your maps and orange squash, whatever you want'. The booklet explained that £3,250,000 had been spent on plant at Cowley to build the car, with special overhead conveyor belts on the line. The staid Ford Anglia would swiftly outsell the Mini, and certainly make more money, but BMC had genuine claims to be an innovative force.

The dealership and after-sales network had also been smartened up in anticipation of the launch. The customer received reassurance that spare parts would be available everywhere. The common 'crash list', containing ninety-two parts such as windscreen glass and bumpers, sump and gearbox casings, all those items 'which are easily damaged or pilferable in transit', would be swiftly obtainable from every BMC dealer, all of whom had attended special courses to improve service. 'If, for instance, an owner, blinded by pride in his new possession, drives out of the showroom and into a lorry, his car can be

*whipped back again and made ready for the road once more
before his tears of chagrin have had time to dry.'*

From the Driver's Handbook
Starter Switch: The starter switch is controlled by the circular
black knob positioned on the floor just forward of the driver's
seat. Push the knob smartly downwards to operate the starter
and release it immediately the engine fires. Should the engine
fail to start, wait until the crankshaft comes to rest before oper-
ating the starter again.

Choke or mixture control: To enrich the mixture and to assist
starting when the engine is cold, pull out the control knob positioned to the right of the control panel. On no account should
the engine be run for any length of time with the knob pulled
fully out.

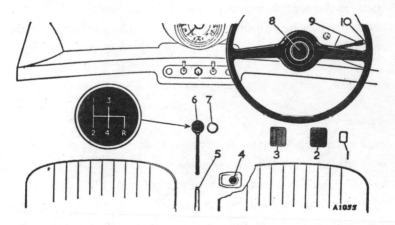

The Driver's Handbook: No. 4 is the starter button.

Running In: The treatment given to a new vehicle will have an important bearing on its subsequent life, and engine speeds during this early period must be limited. The following instructions should be strictly adhered to:

During the first 500 miles DO NOT exceed 45 mph; DO NOT drive at full throttle in any gear; DO NOT allow the engine to labour in any gear. BUT research has proved that the practice of warming up an engine by allowing it to idle slowly is definitely harmful. Allowing the engine to work slowly in a cold state leads to excessive cylinder wear, and far less damage is done by driving the car straight on the road from cold than by letting the engine idle slowly in the garage.

Seat Belts: Seat belts for the front seats (Part No. 22A754) are available from BMC Service Ltd. Adjust the short belt until the attached buckle is located just in front of the hip. The upper part of the long belt passes diagonally across the chest; the lower part returns around the waist to the door sill attachment point. The belt is fastened by pushing the buckle tongue into

A4892A

the buckle until a positive click is heard. To release the buckle, lift the buckle flap to approximately 90 degrees and exert gentle forward pressure on the belt at the same time.

The fitting of seat belts to the car should only be carried out by an authorised dealer or distributor.

Luggage Compartment: Additional luggage-carrying capacity is provided by making use of the lid in the open position, and to allow this the rear number plate is hinged.

Use the lid for carrying bulky rather than heavy articles.

PERIODICAL SERVICE ATTENTION
Daily:
Check oil level. Top up if necessary. Check water level in radiator.

Weekly:
Test tyre pressure and regulate if necessary. Check battery level and top up if necessary.

3,000 Miles or Three Months:
Top up carburettor piston damper. Clutch – check level of fluid in supply tank. Check brakes and adjust if necessary. Lubricate all grease nipples.

6,000 Miles or Six Months:
Check fan belt tension, valve rocker clearances, distributor contact points, functioning of automatic advance and retard mechanism. Clean and adjust sparking plugs. Check front wheel alignment. Check tightness of all nuts and bolts on universal joints. Lubricate dynamo bearing.

12,000 Miles or 12 Months:
Drain engine/transmission oil and refill with fresh oil.
Lubricate water pump, sparingly.

Roy Davies

When you're totally immersed in a factory, most vehicles are just another vehicle so you don't get the buzz on it until it's launched. Once it's out there then you can begin to see.

One of the fascinating things I always remembered with the Mini, they'd no sooner bought it, the people, than they wanted to make it go faster than it had ever been designed to do, and, I mean, one of the very frail pieces of the Mini which we quickly discovered was the drive shaft arrangement, which was basically a rubber spider encased with two halves of metal. And the instruction on the assembly line when you were sandwiching them together was 'Don't have more than a quarter of an inch of thread showing through the nut, otherwise you've pulled it too tight.' Well the Mini suffered quite dreadfully from oil leaks in the early days. Being a totally new engine concept, and probably not being machined to quite the requirement it should have been, the drive shaft joints took an awful hammering with the oil, and of course the more the oil got on the rubber joints, the softer they became. All the go-fast boys used to love to put something bigger under the bonnet, but of course they didn't realise that the first corner they came to was really going to take the load, and it wasn't unusual for some of these people to put two-litre engines in, and go flying off down the equivalent of a Silverstone circuit, and when they came to the first corner, the joints all flew to pieces, because they couldn't cope with the sheer power being driven through them.

By and large we were building them with green inexperienced labour, and a Mini was a complex vehicle to put together, particularly bolting all the subframes. Peter Tothill and Tony Monk had got rid of dozens and dozens of bolts, but putting them in at the back end was still really difficult, and there were a lot of cross-threaded studs. There was quite a long time before you get the [modification] information rolling back, because you're pushing – well, at the time we were pushing out twelve hundred cars a week. They were going out and then coming back, and then you've got to talk to people and get the problems resolved.

A year after launch, BMC published a booklet celebrating the Mini's achievements. Many months before John Cooper began tuning their engines, the Austin Seven and Mini-Minor both won European small car rallies in 1959 and 1960. The car also won prizes for its looks: in the Eighth Southsea Concours d'Elegance, Jane Collis and her highly polished Mini-Minor won the Class Five category, for cars entered and driven by lady members.

There were also details of a trip taken by two writers for The Autocar, *Peter Riviere and Ronald 'Steady' Barker. They had driven almost 8,200 miles around the Mediterranean in a Morris Mini, and on their return to London they stripped down the car to examine wear. 'The drivers were astonished to find little evidence of their hard work' driving through Le Mans,*

Bordeaux, Madrid, Fez, Beirut, Belgrade and Nuremberg.
Overall fuel consumption was 36 mpg; the best day's run, 506
miles from Benghazi to Misurata, achieved an average 53 mph;
in Libya they managed 82 miles at 66 mph.

Ronald 'Steady' Barker

I had joined *The Autocar* in 1955. I suppose my immediate
reaction to the Mini was its extraordinary manoeuvrability.
And its safety. When you were faced with an accident situation,
instead of ploughing into the other vehicle, you drove round it.
This later proved very important in rallying.

The Autocar's technical editor was a man called Harry
Mundy. He had designed the V12 Jaguar engine. He said that
the Mini 'is not a world car', which of course it wasn't, not in
the same way as the Beetle. The Beetle always made money
and got popular in America, but the Mini never did get popular in America for obvious reasons – the Americans are too big
for it.

We actually left the country on the day the Mini was
launched, waved off from the Festival Hall by Jack Brabham in
a wheelchair. The Mini had not been seen anywhere.

I got arrested in Turkey for something I hadn't done, connected with how much money I had. I had to stand trial twice in a
day. And there was a war going on in Algeria, great trouble
getting there even though we had all the visas, and we saw people running up and down mountains with machine guns. People
were absolutely fascinated by the car.

When we got back the car was totally stripped, right down to
the last nuts and bolts, to see what state it was in, and then it

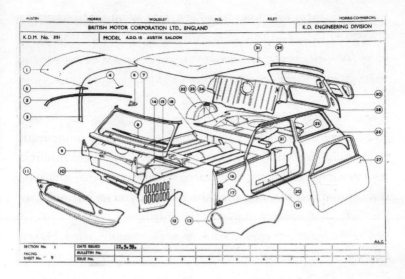

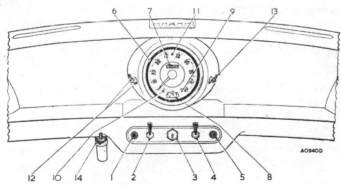

The instruments and switches (Standard, Van, and Piek-up)

1. Heater control.
2. Wiper switch.
3. Ignition switch.
4. Light switch.
5. Choke control.
6. Speedometer.
7. Total distance recorder.
8. Fuel gauge.
9. Ignition warning light.
10. Oil pressure warning light.
11. Headlight main-beam warning light.

12. Parcel shelf light and switch (instrument light and switch—L.H.D. models).
13. Instrument panel light switch (parcel shelf light switch—L.H.D. models).
14. Windshield washer control.

was all put together again and somebody crashed it and wrote it off. We had several problems on our trip, and all of these were pooh-poohed by Issigonis. 'Oh, of course we knew about that!' But they just hadn't fucking done anything about it.

The same BMC booklet included letters of congratulation sent to Leonard Lord from contented motorists, including one from Spike Milligan. 'I am so impressed with my new Mini-Minor, I simply have to write to tell you what splendid value it is for the money, and it is the only car for city travellers. Its parking and performance make it the most notable car of its class in the world. You have my permission to use my name in any publicity for this motor-car. Gratis.'

The tiny tarmac footprint: squeezing in between two Morris Minors

Then there was the export drive. The Mini was acclaimed as a hit in Canada, New Zealand, Malaya and Sweden. Demand was so high that soon BMC claimed it would produce four hundred thousand cars a year. It was overambitious: the millionth Mini was not made until February 1965.

Roy Davies

I don't want to be too hard on the car, but when we went on our Swedish adventure in January 1961 . . .

It was at the invitation of Smith Industries [radiator manufacturers]. It had an environmental chamber at Witney, Oxfordshire, which at the time was the most sophisticated in the country, and every manufacturer who wanted it would go into that chamber and get his results. They wanted to make sure those results were correct, so when you were trying to compare one car with another or had put a bigger heater in, they always took everything back to minus 32 degrees Fahrenheit to give a comparison. Every five years Smiths went to a country where they could get a consistent figure to see if the results were the same. So that was the reason for going to Sweden.

I flew to Stockholm in a BEA Viking Airliner, my first flight, my first visit abroad. We then drove two hundred miles northeast in the Morris Oxford contact vehicle to Malung, out in the wilds, deep snow and frequent heavy snow showers, a centre of the fur trade. The local population hadn't seen a Mini before, and they were envious of its performance. It did prove to be the best vehicle for driving in those conditions, motoring along the main roads at up to 50–60 mph. On the side roads you trusted to luck.

But what also used to happen was that you would come out

in the morning to do your test, earlyish in the morning, it's cold. You'd drive up to the first crossroads; you couldn't see where you were going because the screen was not demisting at all, so you'd make yourself a little six-inch hole to look through, which was great for looking that way, but not great for looking that way. So of course all the windows were absolutely saturated and you'd go to pull the windows but they'd frozen up, so they wouldn't move. So you'd think, 'Next move, I'll open the door.' But of course the door was frozen up and when you pulled the chain the door nut snapped so you were then trapped in this car.

In 1961, The Morris Super 850 promised 'Now! Mini-motoring in maximum luxury.' (Again a glamorous couple in evening dress, but now photographs replaced drawings.) 'New Comfort. New Glamour. Note the attractive new door chrome finish and the bright modern radiator grille. You'll love the amazingly spacious interior with new splendid furnishings – new design seats, new attractive appointments everywhere.

'It's the perfect car for the fastidious.' There was a new reading light as well, new ashtrays for driver and rear passengers, new door handles (gone were the plastic-coated wire pull-down strings, replaced by a metal lever), two sun visors, one with mirror, and a new oval instrument cluster including an oil gauge and ammeter.

When demand improved in the early sixties, the Mini was being made at the rate of about six hundred a shift. When more were needed they started making them in the north works as well, mixed with the old Oxfords. About seven Minis an hour

*were sprinkled into this other, much more complex, build. And
some of the photos show that, for a while at least, the production lines combined both Morris Minis and Austin ones.
Between 1961 and 1968 over 115,000 Austins were built at
Oxford.*

*The Mini van, with no rear seats or windows, first appeared
in 1960. No purchase tax was required on this commercial
vehicle, and the price came down to £350. Very slowly, this
model began to replace bicycles in the Cowley and Longbridge
car parks. For the first time, the Mini was within reach of those
who were making it.*

5 'My God, what a car,' thought John Cooper

Roy Davies

After production was in full swing we got lots of problems coming back to us from the field. One interesting one was the crabs. They were Minis that would run down the road sideways. Literally, they would be running – you'd start at the top of a hill, hold your arm on the steering wheel, you'd finish up in the kerb at the bottom. So you measured your bodies, you measured your subframes, everything's within tolerance. You went out and got one of those sophisticated devices that measured all the wheels at once . . . It turned out that the frame was marginally twisted at the back, because the body panel was slightly out, and the frame was slightly out, so you were just running slowly down the road, offset. So with that you had to go all the way back through, talk to your subframe people, sort them out, sort the body panel people out, then you were OK.

We got big contracts with ministries, and a huge contract came from the NHS for what was called the Midwives' Mini. And you were very much in the hands of these various authorities, because there were Minis for the navy, Minis for the RAF, Minis for the army, Minis for the NHS, Minis for the police, Minis for the Post Office, and they'd all got edicts of what you were to do with them. The navy regarded it as a ship: all the

Opposite: The end of the production line at Cowley

subframes had to have underseal all the way over them before they went in. Which made it impossible to put the bolts in, because they were all filled up with underseal. The RAF was just quite happy for the underseal to go on the bottom of it when it was fitted.

Frank Baker was the man responsible; he was the ministry inspector, and he was told that with NHS Minis you must be able to take them straight off the track, drive up the road, hit the brakes full on at 30 mph and they mustn't deviate. Which is almost an impossibility.

The Mini introduced for the first time a peculiar device in the back brakes called a brake limiting valve, which cut the pressure off at a certain point, so no matter how much more you pushed, you weren't putting any more braking into the back, you were putting it all into the front. And the way we solved the midwives' test, we played about with the setting of the valve at the back and made it cut off even earlier, because really the brakes at the back of the Mini never really did anything. So all the way through the life of the car there were always two brake valves available in service, the normal one and the midwives' one.

You were constantly faced with this piecework push. You kept pushing cars out that weren't ready for sale. You had to keep repairing them, so the real drive on the quality front was to get proper right-first-time vehicles, which meant you were addressing all sorts of problems, which weren't the problem of the designer but which were the problem of variability. When you bring hundreds of wheels in, some of the wheels are not going to be right and so you would try and narrow the variability down as far as possible. It's no good going out on to a manufacturing floor if they're having problems with fitting something and saying, 'Get on with it.' Because they don't get on with it. The piecework boys just roll it past you.

So, for instance, to give you an example of a nonsense. We'd been building them for quite some time and suddenly they're all coming down the track with the front suspension shock absorber not attached. All it turned out to be, there was one awkward person who joined the build area. Shock absorbers are about this long, but they can be that long, and that long [he stretches his thumb and forefinger several inches apart], because there's tolerances in them. Everybody had accepted that and they'd pop the shockers on and if it wasn't quite coming on to the top bolt then a knee went on to it and up they went. But this man on the line wasn't going to do it any more.

'I'm not paid to put a knee under; I want piecework money for putting a knee under.'

'Well, you can't have more piecework money because you don't have to do it on them all.'

And that's the sort of thing about which you couldn't say, 'Issigonis, you've designed the shock absorbers wrong.'

It was intriguing the rate of progress that the accessory people made in making the Mini better. Peter Ustinov described the Mini gear lever as a paintbrush. Well, we used to describe it as a pudding stick, because it stirred round and round and round and with a bit of luck you found four gears in it. It was the accessory people who brought in the little extension that made a proper gear lever of it. It may have been a couple of years or more before one actually came in on to the production line.

Lord Snowdon

With the first Mini I had I did various alterations. I put in the first safety headrest that I designed, and that [Issigonis] didn't mind. And then we made the accelerator pedal a bit larger. Then I found someone who put in a wind-down window, in a mews in London, and I thought it was much nicer than those slide windows. I sent it back to Alec for something else, and it was promptly returned with the slide windows put back in. He didn't like that at all – he loathed gimmicks.

Laurence Pomeroy, *The Mini Story*

The sliding front windows are an integral part of the ADO15 concept, but there have been a few who have clamoured for the wind-down type. Such a change would reduce elbow room, increase weight and cost, and encroach on the internal door wells, which are such a joy for those who use their car as a perambulating library, kitchen and a study for geography. Mini pockets, in fact, like the deep-freeze container, should be thoroughly cleaned out every six months so that the penny bar of

Sir Donald Stokes, Sir Alec Issigonis and Lord Snowdon argue the finer points of sliding windows in 1971

Fry's Cream which today costs sixpence, the Blackwell calculator used in last winter's rally, and the *Michelin Guide* which was so useful on last summer's holiday are not imprisoned at the bottom of a very deep well.

In January 1963, Morris Motors printed a new brochure.

There was now a turquoise car and a lime green. You could drive on white-wall tyres, which in 1959 had only been available on the Morris version. Marketing had invented the concept of Mini-miles. 'Mini-miles! The new yardstick for modern motoring. Mini-miles are cheaper, faster, brighter. More enjoyable.

'So pretty . . . yet so practical.'

Seem shorter too. Mini-miles are so cheap – less than 2d each, including petrol, oil, tyres and servicing! This works out at less than 1/2d per person per Mini-mile with four up. Mini-miles end in simplest-ever parking. An 11 ft 6 in gap and you're in. Without a care in the world.'

These were now 'GAY happy-go-lively cars . . . catching today's mood. Gay's the word for these great-hearted little marvels. They're so pretty . . . yet so practical.

'These amazing Mini-Minors drive like a dream. Corner like cats. Watch them at speed . . . with four weighty adults. At the lights – rarin' to go. Now. Look at the people who drive them. They're "up to the minute". They're going places fast.'

There were also two new models: the Mini Traveller and the Mini Cooper. The Traveller promised extra roominess with rear-opening doors. Folding rear seats meant there was still more room. The Cooper appealed to the performance-minded. 'If you are a real get-up-and-go enthusiast, if you like fast getaways – here's your car!' It had twin carbs, 997 cc, disc brakes and 'a lot more besides'.

John Cooper

I remember once when we were at a meeting with the press, when they were launching one of the Mini Coopers, and one of the press people said, 'When are you going to put more comfortable seats in the Mini?' And [Issigonis] said, 'Well, you've got to be uncomfortable to keep awake.'

Alex Moulton

The key thing one must constantly remember is this: it wasn't that successful on the market, but it was successful in the press. It took a while. The relevant people at the time were show people, and it became a smarty thing from that point of view. And then of course Snowdon. And Daniel Richmond of Downton Motors – he was a brilliant man in tuning the engine – ensured that quite early on there were Minis running doing a hundred miles per hour.

I think it was Steady Barker the journalist who had taken one of Downton's hot Minis to show it to him. Issigonis was surprised, but I think it was the provocation of John Cooper that changed things. It wasn't until John Cooper had tried the Mini and thought, 'My God, what a car.' I remember I was going up the M1 [opened three months after the Mini launch in 1959] in my SV Bentley, running about eighty miles per hour, and there was a bloody Mini behind me. I thought, 'Funny thing.' It turned off, and it was John Cooper taking his hot Mini to show it to Harriman, to persuade him to make a special one. I think Harriman was forced into it rather, against his natural inclination, which was just to make a nice small car. And then the Monte Carlo rallies just wiped the floor with everyone. I didn't have a Mini myself at first – I was quite late into it. It was only during the second Monte Carlo Rally in 1965, when Harriman said, 'I think we're winning,' and Issigonis and myself went down there. That story about de Gaulle saying to customs, 'Make it difficult for the Mini spares to go down' – all that's quite true. [Paddy Hopkirk won the 1964 Monte Carlo Rally in a Mini Cooper S, and Timo Mäkinen won the second in a Mini Cooper S in treacherous conditions the following year.]

The 1964 Mini Cooper S driven by Timo Mäkinen to victory in the 1965 Monte Carlo Rally

The Mini 1000 on a new motorway.

Then I started owning one myself and I thought, 'What a fantastically wonderful thing it is.'

By 1964, with almost six thousand Minis being made each week, engineers had finally worked out how to get Hydrolastic suspension into the car, and it was installed just before the Monte Carlo triumphs.

Sales brochure, 1964
A cushion of fluid between you and the road. Just that! Hydrolastic suspension absorbs jolts, jars and shock with a cushion of fluid. Interconnecting front and rear suspension units, using anti-rust and anti-freeze fluid as a damping medium, automatically compensate for uneven conditions – both forward and laterally. An advance on all other suspension systems because of its basic simplicity, it has no wearing parts, no glands to leak. No maintenance is needed.

Hydrolastic was introduced on the Morris 1100 at launch in 1962, another big hit for Issigonis and Moulton; in its various badge derivations and engine sizes, the car soon outsold the Mini to became the UK's bestselling car of the sixties.

British Pathé News, 8 February 1965
What a day in the life of the British Motor Industry! The millionth Mini came off the production line, a landmark indeed in the history of the automobile world. [Film of Longbridge assembly line, with grinning workers.] To the works' manager and man whose brainchild met with astounding success,

Not much changed: Issigonis with the first Mini and a later model outside a new Longbridge assembly building

An early Mini adventure on a jet plane

designer Alec Issigonis, and all the thousands who work at Mini plants, unstinting congratulations are due. More than a third of that million have been exported in the last five years. Car number 1 million was driven by Issigonis himself. [Issigonis climbs into white car with a big sign on the roof. He pulls open the sliding window before driving off the line with a bumpy start.]

A competition for the millionth car was open to all who ordered or took delivery of any BMC model last December. The winner was a Liverpool ship's pilot, Peter James, here with his wife to receive it. [Clip of smiling photogenic couple being given keys.] Good for him, and roll on the second million!

In 1966, the new 'Cowley Colours' were the first clear influence of fashion on the cars. Issigonis was not a fan of the new two-tone look, not least because the names of the colours again revealed the hand of the marketing department. Buyers of the Mini-Minor Standard, Deluxe and Traveller now had the possibility of Fiesta Yellow, Tartan Red, Surf Blue, Smoke Grey, Old English White and Almond Green.

For the Super Mini-Minor and Mini Cooper there was the possibility of having these colours for the body or the roof. The increased choice also signalled the positioning of the Mini as a car you could design yourself, the driver as individual (there may be a million of the things out there, but none of them quite like mine . . .).

In 1967, the Mark II Minis offered new grilles, optional automatic transmission, new badges, an improved turning circle, a larger rear window, redesigned light clusters, leather-

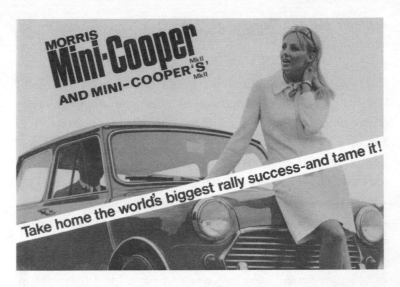

MORRIS **Mini·Cooper** MkII, AND MINI-COOPER'S' MkII

Take home the world's biggest rally success–and tame it!

cloth seats and a speedometer now calibrated in kmh and mph. It was a modern, fresh, European feel. The new station on the optional radio was Radio 1, with Tony Blackburn playing 'Flowers in the Rain'.

Eddie Cummings

It wasn't just at the beginning – they never made any profit on the Mini. If the quality had been better they could probably have charged more for it.

Eddie Cummings, in his mid-eighties, is sitting with his wife Jean in the living room of their spacious and comfortable flat in the Summertown area of Oxford. Jean still has a part-time job at Keble College a few minutes' drive away, where she helps serve students their breakfast. Their memories are being stirred

MINI

by their son Ian Cummings, himself a Cowley worker for almost forty years.

They did skimp on material. It tended to be cheap but not nasty. The trim wasn't very good, it was like sitting on a board almost. There was always penny-pinching over the components. You know, 'Can I get that for a couple of pence cheaper?' And that couple of pence made all the difference between something that was reasonably good and something that was basically rubbish.

I was born and bred in Oxford, and in the navy I went all around the world, went around the Antarctic for a couple of years. I started at Morris Motors as a lowly pay clerk in the wages department, then I moved into an area concerned with production control, which is when I first worked on the Mini when it started. They used to race it round the factory – it was almost impossible to turn over.

I ended up as chief safety engineer. I was responsible for safety throughout the factory, dealing with the factory inspector and the insurance people. I often had to go to London, to the magistrates' courts when people were claiming damages for accidents at work. Some accidents were quite serious, but we tried to keep the amount paid out as low as possible. From being quite a cushy job it became quite a nasty one at times, and we had a lot of bolshie unionists.

We had a couple of fatalities of contractors working on site. A board in the roof girders collapsed, and a man came down about thirty feet on a concrete floor and was killed straight away. And the other one, there was a pole that was supposed to be a safety barrier, and a driver at night came rushing up this tunnel, and the pole went straight through his windscreen and killed him. We never got prosecuted and I was very surprised. We had it repositioned after that.

Compared to Pressed Steel next door, Morris's wages were quite low, but compared to the economy generally they were pretty high. When people were earning £10 a week in the general economy, maybe a man at Morris's was earning £30.

When I started in 1955, Morris Motors employed about ten thousand people, and when I left [in 1987] they were producing the same number of cars with four thousand.

Jean Cummings
My father worked at Pressed Steel Fisher.

Eddie Cummings
It was like walking into Dante's *Inferno*. If you went into Pressed Steel where her father worked, spot welding, it had its own atmosphere, it was just a haze, there was virtually no extraction, it was filthy. It was a real sweatshop. They always said that on a bus you could tell the difference between a Pressed Steel worker and a Morris worker – the Pressed Steel worker was asleep.

Jean Cummings
I never thought I'd be working at the plant as well. I went up there purely for money. I had four children, and they were all pretty clever and got their scholarships and needed uniforms, we had a mortgage, we needed holidays and my husband had just got a car.

I used to go off to work as soon as the children were ready for bed. I was in the trim shop, I did the seats for the Mini. But I was never any good on the sewing machines, I could never keep a straight line. They didn't ask you whether you were good at sewing, they just said, 'You've got to stitch round there, and you push it round as you go along.' I thought, 'Easy.' I was doing the back, and I had to sew the Rexine on to wadding. It was rubbish – you try to stitch Rexine, it splits easily, it was awful. The inspector used to throw my work back at me, and I wasn't earning any money, because the quicker you got them out the more money you earned.

So they put me on something called Progress, which I was very happy about – I delivered the materials to each person on the line. When you saw them running short of material, we had a truck thing with the material on and you went to each person. I earned better money. But I didn't just have the Mini, I was also going over to the other side where they had the leather for the Morris Oxfords.

Then they got rid of the twilight shift and I lost my job. We had an awful row with the unions, because we were paying union fees, and I asked for some sort of compensation, but they didn't want to know.

Eddie Cummings

They always used to say that Morris's made the profits and Austin spent them. When I first went there I think it was quite a happy place, but towards the end people became a bit despondent. Once Nuffield left, he passed the reins of power to Austin, and we always thought we were second best.

I've never been in the new MINI.

Ian Cummings

He drives a Nissan now.

Jean Cummings
I remember there used to be a little Morris garage in Cornmarket Street in the centre of Oxford. There was a hotel next to it, and I can remember my father taking me there when I was about seven. My father drove Lord Nuffield around several times. Once, I was sitting in this very posh car, and it was Lord Nuffield's.

In May 1961, at the age of eighty-three, Lord Nuffield was interviewed by the Daily Express. *He was long through with the car business. In fact, he was through with almost every business apart from giving away his money. His health was failing, he had no heirs, he had been lonely since the death of his wife two years before, he was no longer able to drive his Wolseley Eight. The interview was a rare event: he disliked the press and being photographed, although it wasn't publicity itself he shunned; his charitable donations were rarely anonymous.*

His giving began in the thirties, with £2 million toward the Oxford Medical School Trust, £1 million to Nuffield College and £1,650,000 for the Nuffield Trust for the forces. During the war he provided his Nuffield Foundation with £4.8 million for the advancement of healthcare, the training of medical staff and the care of the aged. Alongside many other donations to hospitals, charities and nursing homes, he supported boy scouts' funds, the Children's Safety First Campaign, the Territorial Army Sports Board and the Kipling Memorial Fund. He also built some of the first iron lungs at his factory and gave them away to any hospital who asked for one.

Nearing the end: Lord Nuffield and a Morris Minor

In his lifetime Morris would give away about £27 million; rather this, he believed, than the furnishing of death duties. But giving was troublesome, he told the Express. *It invited letters from causes he didn't consider worthy. Individuals would write to him with sob stories, and he tried to read them all personally and reply, especially to the ones he found complimentary. But now he said he was sorry, but he couldn't possibly deal with them all.*

Within hours of the interview appearing in print, letter boxes were filling up throughout the country. Approximately three hundred letters arrived within a week, and many are filed away in the

Nuffield Papers at the Heritage Motor Centre archive in Gaydon, Warwickshire. The file is called 'Begs – not acknowledged'.

From Leslie G. Luker, Principal of the London College of Pharmacy and Chemistry for Women

28.5.1961

Dear Lord Nuffield,

I have just read with great interest of your interview with Mr Donald Gomery of the *Daily Express*, and I hope this will offset some of the misery you experience after announcements of your benefactions. First, please allow me to congratulate you on looking so fit at 83 years of age. Although nearly thirty years younger, I envy you in this. Like you I have suffered from sciatica, but this was due to a motor smash in my youth.

Many of your benefactions have been in organisations close to my own heart, particularly in the case of St Dunstan's (I have been temporarily blind twice owing to an explosion).

I would like to tell you something of our own work at the college. Founded in 1892 by the late Dr St Aubyn Farrer, of Harley Street, a physician to the household of King Edward VII, it was well equipped and maintained as a private charity or hobby, to train ladies of birth, but often of small means, as hospital dispensers.

Unfortunately, Dr Farrer was blinded in a car smash in 1932 . . . We would like to be able to offer some scholarships or bursaries to suitable London and Middlesex girls who are anxious to serve the community.

MINI

From Mrs Edith G. U. Byrne, Peacehaven, Sussex

28 May 61

Dear Sir,

Would you lend me one hundred pound please. I have discussed this with my four children who have agreed to pay you back when they get to work. The reason I am asking you is this, my garden is such a mess, both my son and myself are Bronchial Asthmatics, I can't work and I am receiving National Assistance. You may think this is quite a wrong thing to ask money for and a large amount to spend on a garden, the retaining wall collapsed and that will take a large part of it. Yours truly.

From Dorothy G. Baillie, Cliftonville, Margate, Kent

24 May 1961

Dear Lord Nuffield,

My husband had to have his right leg amputated through being shot in the ankle during the First World War. He served with the London Scottish, and through this he had to retire from his business as Manager with a large Organization as he could not get about as quickly as he previously used to.

When he retired the firm were wonderfully good to him and gave him the Morris 16 he used for business purposes, but like all good things it had its day, and we had to dispose of it as it is no more road worthy and costing far too much money for its upkeep. This is – I must admit – a very great handicap to my husband, who cannot even enjoy a bus trip on account of the cramped position in these vehicles, and therefore I am not able to get him out and about because he cannot walk very far and I just hate to see him missing his outings, and that is why I am writing to you, to ask if you have – among your many possessions – a car you have finished with, I would willingly come along to any place you might name to pick same up, or I would welcome you, so very much, if you are ever in this district.

Believe me,

Yours faithfully.

The next time Lord Nuffield received so many letters, he wasn't there to not acknowledge them.

Daimlers and Minis at a Worthing dealership

**From the Managing Director, The Wicliffe Motor Co. Ltd,
Morris Distributors, Stroud**

August 22nd 1963

It was with real regret that we heard the announcement of
Lord Nuffield's death this morning . . . It is impossible that he
should not be remembered by generations to come, and certainly will never be forgotten throughout the Industry to
which he has contributed so greatly.

From W. H. T. Powell, Powell's Garage (Abergavenny) Ltd

To: the Personal Secretary, The Late Lord Nuffield

I sympathise with you at this sorrowful time . . . It may not be
of much interest to you, but my father (now deceased) and his
Lordship were great friends and friendly competitors at one
time, when in the nineties we were manufacturers of the
'Powella' Cycles and MotorCycles. I personally recall the
many friendly chats that he and I have had together since we
were appointed Morris Dealers in 1914.

**From D. H. Phillips, Managing Director of Bristol Motor
Company, Ashton Gate, Bristol**

We are all very deeply grieved at the passing of Lord Nuffield
and offer our very sincere sympathy to that which will be
expressed by the countless number of people who have been
helped in so many diverse ways for all that he has done for
mankind the world over.

In the week these letters arrived, the Mini celebrated its fourth birthday. The car was still a unique proposition on British roads, but it was not yet a popular choice. Its engineers were still working on problems and improvements. It was still losing money, and perhaps this was why Nuffield did not consider it a great success. But then Twiggy climbed into one, and Peter Sellers and the Beatles and the whole of Swinging London tried to get in after her, and the Cooper version of the car kept winning the rallies, and three of them – red, white and blue – pulled off a bank heist in the centre of Turin. The money stolen in The Italian Job *(1969) came from the Fiat car company. 'Four million dollars – through a traffic jam . . .'*

Loaded with bullion in *The Italian Job*; Mary Quant in her own swinging Mini

But to the people in Cowley, the Mini wasn't the Mini anymore. The car left the plant in 1968 after opposition to a night shift and the planned introduction (the following year) of the Maxi, the last car to be designed by Alec Issigonis. About six hundred thousand Minis had been made at Cowley, but now the British production of the car continued exclusively at Longbridge, where it suffered from neglect and indecision.

When Mini returned to Oxford thirty-two years later, everything would look different – the car, its owners, its price, the plant, the way it was spelt, and the way it was made.

PART TWO

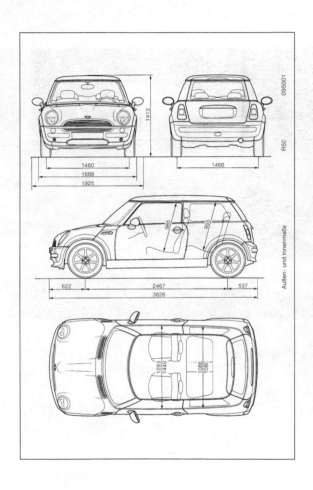

An early glimpse of the new MINI body at Cowley in 1999

6 'And now the future rested on one model,' said Paul Chantry

Paul Chantry (retired deputy managing director)
On 16 March 2000 we were told that a big announcement was coming, and the management of the plant were told to get into the personnel director's office and await further instructions.

This had all been kept really quiet.

First thing in the morning we had our initial production meetings, then we all turned up, and waited. There were a few phonecalls from the Rover Group headquarters in Warwick, and then more waiting. In the end a man turns up, I think he was a Group 4 security man in a green uniform, with a bundle of hand-outs, and he put the bundle in the middle of the personnel director's desk.

And he waited.

Then a phonecall came: 'Do not open this.'

So we sat there and had lunch. And then we still sat there. In the middle of the afternoon, this Group 4 guy arrives again and took the package away. We're all bemused, and the personnel director is ringing up Warwick and saying, 'What's going on?'

In the end we said, 'Look, all the guys at the plant will be going home soon, and the last thing you ever do is announce this through the press . . .' So then we got a faxed copy of the announcement.

We had one copy of the fax, and we had to photocopy it

however many times – maybe there were two thousand people. It said that Rover Group was going to be split up. It had been agreed that Land Rover would be sold to Ford. We would continue to make the Rover 75. And it said that MINI was coming to Oxford.

Paul Chantry is in the bar of a hotel opposite Nuffield College. He retired from Plant Oxford (as BMW officials now like to call it) in March 2007, aged sixty. The public-relations people had given him the title of deputy director of the MINI plant, because he had often stepped in when the German director had been called back to Munich. He had read physics at Oxford, worked for English Electric at Rugby, and then moved into aerospace, a company that became part of Marconi. He worked on airborne computers, including the Concorde.

He tells me of the three Minis in his life. He bought an all-metal Mini Traveller when he had a young child, folding the Silver Cross pram into the back and going camping. Then there was a good-looking 1965 model, almond green with a creamy white roof, which he sparked up with really stiff shock absorbers, a remote gear change and leather steering wheel. And in 1976 he bought new for the first time, but the car was written off in an accident after three months.

We had seen the prototypes of the new MINI on paper and film. But a day after the announcement, my then boss, a guy who eventually went out to China to run BMW's new plant, he got a MINI down here. And so four directors got in and we drove it around.

Ian Cummings (process improvement manager)
With the MINI, BMW realised they had to have a car in its fleet
to meet the emission standards – all these European regulations. The BMW 1-Series was years away: we had to have the
MINI – what's the best route to have it built somewhere? There
was no way that as a stand-alone car they could build it at
Longbridge. The scale of the factory, and all that went with it,
too big, it would have been in the middle of all the hived-off
Rovers, and BMW wanted to be rid of all that. But this factory
at Cowley, it was almost a custom-fit. You lucky people at
Oxford . . .

Chris Bond (deputy plant convenor)
The threat of closure was still on us. When the news broke
that they were going to keep Oxford and the MINI was coming down, everybody was so elated. They thought the car was
wonderful.

Frank Stephenson (chief designer of the new MINI)
The MINI Cooper is not a retro design car, but an evolution of
the original. It has the genes and many of the characteristics of
its predecessor, but is larger, more powerful, more muscular
and more exciting than its predecessor.

Paul Chantry
We were impressed with the space – four guys got in there,
and none of us were small. But we were concerned about how

difficult it was going to be to make. I could see that we were going to have to make it pretty accurately. There's a hell of a lot packed into a very small space.

The new model: an early Frank Stephenson sketch

The car had many fathers: it started off as a Rover project, and then a BMW project, then just before it launched it was back to being a Rover project with some BMW input. And then when Rover was sold, BMW took the project back, and we had to achieve the true BMW engineering standard, changing fixings and materials to get to the BMW expectations of durability and corrosion protection. There was a significant change to the rear end to improve the crash impact, but that was the only thing. Product and development engineers, they're very, what's the word? Arrogant. Their way is the only way, and they would not release a drawing if it didn't comply with the standards.

But no one was really sure about it. We began to be a little bit more convinced when we got some cars built down here. You started to throw it around the roads around Oxford, and you realised it was fun, but also a very comfortable and quiet car.

Pat Nolan (engineering manager for framing and bodysides process planning)
I was at Longbridge, and there were lots of rumours that they were going to bring in a new model. We were working on various joint projects with Cowley, but the MINI wasn't one of them. But then they started doing the clays of the new MINI at Gaydon [Warwickshire] in 1996, life-size clay models. There were changes to what it looks like today, some were bigger and some smaller, and they had to do their clinics and decide what to do. We used to go and have a look at them. We thought, 'Well, it's different,' but we couldn't imagine BMW doing it really.

Then they suddenly needed almost all the hands they could get to work on it. It was a big job, we had to clear almost all of West Works in Longbridge, we had a lot of contact with the designers at Gaydon, then we had to go over to Germany almost every other week with the tool manufacturers and put the processes to them. We were also then getting into BMW systems, and how they did the job and the standards they used. A very exciting time.

Pat Nolan arrived at Cowley in 1967, about the same time that the Mini departed. His father had joined ten years before, after leaving the RAF.

The ACV 30 concept car from 1997 (top and middle) and Rover's Spiritual and Spiritual 2 prototypes (bottom) that never made it

Pat Nolan never owned a Mini, but he passed his driving test in one. He began at Cowley in jig and tool design, drawing the tooling demanded by the planners for their prototypes. The cars he remembers most clearly at the start were the Maxi, the Rover P6, the Hillman Minx and the Sunbeam Rapier, and the robots putting together the Montego. He later did the tooling for Rolls-Royce bodies, and at the opposite end the first Korean car – the Hyundai Pony.

Even though we all thought the new MINI was going into Longbridge, the prototypes were built in T Building in Cowley, at the back of where the visitors' centre is now. We were used to the way that Honda built cars, which wasn't as good as the way BMW put cars together. The way the MINI was going together was completely different – a lot more stiffness, a lot more of the integral welding done by robots.

All of the people working on it were very excited. Not just the body, but what was inside there. And when they were being driven and they kept the same road-holding ability as the old Mini . . . the stiffest car in its class, so you don't get any shake or rattle. And the size of it as well – when you put it next to the BMW 3-Series it was as high, and almost as wide, and it wasn't the baby Mini any more. It was really the first prestige small car.

Paul Chantry
But the news was constantly changing. Originally Jon Moulton was going to buy Rover with the Alchemy group. The 75 was still going to be built at Cowley by BMW under licence, and the MINI was going to be squeezed in alongside it. So we started

working on plans, and that lasted for a couple of weeks. Then the alternative plan developed: the Gang of Four with John Towers would form this Phoenix consortium.

We'd spent a lot of money here, we'd already put the new paintshop up for the Rover 75, so Cowley was a far more modern and efficient plant. The Rover 75 had been a success story from the manufacturing point of view. We had a very flexible workforce. So between Longbridge and Cowley – that would have been a pretty clear-cut business case. The general political impression of the plant was more favourable than that of Longbridge. Even with the Rover 600 and 800 we'd improved the productivity of the plant by a hundred per cent, by closing the other side of the road, the old Morris plant. So the Oxford plant was not bad, whereas Longbridge was not good.

So we started working on plans for both cars, and that lasted for a couple of weeks. Then I got a phone call that said, 'Find a way of moving the Rover 75 to Longbridge.' So I went up there with a couple of my colleagues, and over about forty-eight hours we came up with a plan. That's when this plant became MINI.

BMW kept the MINI in the UK because they had buried all this money in the project and they had to do something, and consequently some people thought their hearts weren't really in it. That's why a lot of people took the opportunity to leave when they were given it. I was the only director who stayed. Some went to Bentley, some to Rolls-Royce. But I'd had quite a lot to do with the Germans, I was the one going to Munich getting investment for the Rover 75. I had always enjoyed working with them, so that made me feel comfortable. But nobody else felt comfortable.

Frank Stephenson
Everyone thinks I'm American, but it's pretty much just the accent.

Frank Stephenson spent his first eleven years in Morocco before moving to Istanbul. His father worked for Boeing, and in the early sixties he opened a car dealership in Malaga, where some of the most popular cars were Minis. He jokes that his mother drove a Mini Cooper S, and it was the first car he ever stole. He remembers the go-kart feel: 'You can tame the car instead of the car taming you.'

He is sitting in a conference room at his new employer, McLaren, in a futuristic building designed by Foster and Partners near Woking in Surrey. All around him are classic sports cars and pictures of a victorious Lewis Hamilton.

I finished my last couple of years of high school in Madrid, and my choice then was either to race motorcycles, which I'd fallen in love with, or move down to my father's dealership. I started my racing career, which lasted about six years, and then my father said, 'It's time to get serious, because you're going to burn out when you're thirty. You should either start working or study.'

He moved to the Art Center College of Design in Pasadena, one of the best schools for car design. He was sponsored by Ford, and after completing his course in 1986 he worked at their European headquarters in Cologne. His main project was the Ford Cosworth, with double spoilers. He says it started out with three spoilers, but then finance people got involved.

He moved to BMW in Munich in 1991. He worked on the X5, but then he heard he was wanted to work on something else.

I think the new MINI is one of the great design endeavours of all time. When BMW acquired Rover, one of the brands there that they really felt had a future to it was the Mini, simply because it was the one that had the most heritage, the most impact. So, how to revive it?

They opened up the competition to all the designers in BMW and Munich, the Rover design group in England, BMW in California, and they were also looking for some proposals from Italy. In total there were fourteen or fifteen full-size design proposals. The designers worked in isolation pretty much. It's incredibly expensive to do a new car like that – normally it's maybe three proposals.

Designers from Rover produced the Spiritual and Spiritual 2, two- and four-door cars that resembled a shrunken Renault Scenic. They were light and airy, with rear-wheel drive from an engine located beneath the rear seats. They didn't look like a Mini, or a MINI, although they did measure about 10ft and have a wheel at each corner.

The ACV 30, a far sportier model, also emerged at this time, and appeared publicly in 1997 at the thirtieth anniversary tribute to the Mini victory in the 1967 Monte Carlo Rally. Internally it resembled a stock car, but externally it provided several clues to the masculine and dynamic look of the car Frank Stephenson was building.

The brief was 'Interpret the new MINI as you see it for the twenty-first century'. So each designer had quite a lot of liberty to inject his personality or vision. Obviously we had a general package: the car had to be small, but we weren't limited [by the obligation] to carry over components, because there weren't any to carry over.

The final show, which was in October 1995, at the Heritage Centre in Gaydon, when they had to choose which one it was going to be, was really one of the most incredible presentations you could ever see. Such a wide variety of design to choose from. My approach was a bit obvious and logical, but I don't think anybody else used this approach. The original Mini had so much character built into it, and had such a loyal following and such a critical eye on it. So what I basically did was re-sketch the MINI for 1969, how it would have looked like if it had changed then. Then I re-sketched a design for '79, then I did an '89 sketch, so the car progressively got more modern, and my '99 sketch was the car that I thought, 'That's pretty much what the car would have looked like if it had evolved every ten years, which it didn't.' That '99 sketch was the twenty-first-century MINI.

Gert Hildebrand (chief designer on MINI from 2001)
The design expression of the car – this friendliness, the combination of baby-like proportions, big head, small body, the masculine stance, and the very feminine sculpture – these are the only body archetypes that exist, man, woman and child.

From his office in Munich, Gert Hildebrand says he was six when the Mini was born. Since then he has worked on small cars at

The design team from 2001, with Gert Hildebrand (middle) and Frank Stephenson (far right)

Mitsubishi, Opel and Volkswagen, designing an improved Golf and launching the new Beetle. He joined BMW to work on the MINI a few months after the new model had been unveiled, and he remembers feeling 'jealous, in a positive way' when he first saw Frank Stephenson's design on television and in Car *magazine. He saw in it one thing above all others: a process of natural evolution.*

I think every product you create – a radio or a bottle – transports the soul of the creator. The engineer designs the form which comes out of his hand and his character. I even go so far as to say that an athletic person designs an athletic product and a leptosomic person designs a leptosomic product. This is authenticity. If you are allowed in a company with all these committees, and the personality of yourself is reflected in the product, then it's authentic.

MORRIS / AUSTIN '!

MINI MKⅡ '68

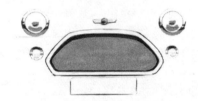

ACV 30 '97

R50 '01

R 56 '06

Frank Stephenson

There were a lot of things when I did that car that were 'impossible', that people immediately said, 'We're not going to be able to do that – the clamshell bonnet, it's such a large piece of sheet metal that we're not going to able to press a bonnet like that.' And the wraparound glass that was extremely wraparound, the tail-lights – those were all things that weren't done on a volume car. But basically the car turned out as the sketch, and we really had to push it to get it there.

On that day in October 1995 we were all waiting for the results. Chris Bangle [head of design at BMW Group] presented all the cars to the board of directors from BMW and Rover Group. He presented full-size models of the car made out of clay, with Dynoc over it, a flexible film on which you apply colour. If you stood back thirty feet you could almost start to believe the car was real. We hadn't started the interiors yet. On that day in October, Chris Bangle said, 'OK, they've selected your car, and it's your job now to take it to production, I want you to go to England and base yourself there in the Rover studio and spend however much time is needed.'

I arrived at Gaydon in November 1996, and they showed me the car, and I thought, 'Well, some things have changed on it,' so we had to try to bring the car back to the original condition as the car that was selected. Just slightly different surfacing and things like that – it was basically the same car, but there's a process called Global Modelling where you can try different things and stretch and pull. The car just had to be re-fine-tuned and brought back slightly into the original look. For the next four years the car is [still] in clay and you fine-tune all the design, all the lines. When you do the feasibility work with engineers, if anything has to change you can change it on the clay model.

We were aiming for 1999/2000. We had BMW engineering and Rover engineering working together on it – they were able to do everything I asked for. It wasn't like I had to jump up and down and scream and holler.

The computer-generated cross-section from 2001

The idea for example of the headlights being tied in to the bonnet and going up with the bonnet – that was a challenge, because only one car had done that before, the Fiat Coupé, and that opened up other feasibility issues: what do we do with the indicators that always have to be pointing forward? With wraparound glass we pushed glass technology to the limit to get that amount of curvature in the front glass. The main thing for engineering was to keep the weight down.

When did I first drive the car? The experience I most remember was not actually driving it, but I drove alongside it. I drove in another car to see what it looked like. Suddenly it turns into a non-static model, so you can see what it looks like in different scale on the road, with other cars next to it.

The first experience I can remember of physically being in the car was with Rauno Aaltonen [the 1960s Monte Carlo Rally champion in a Mini Cooper S] at a test track in Munich. I didn't know who he was until he came up to me and said, 'I'd like to show you what I think of this car.' He was limping because he'd had a snowmobile accident. He got in the car and drove towards these snow cones. We had a set speed that you could come in at, and he was much much faster than that. He did that Scandinavian flick, where they back the car into the corner. Then he got out and said, 'I just wanted to tell you that this is the best-handling front-wheel car that I've ever been in.'

I said, 'Yes, but who are you? Because nobody drives like that.'

He said, 'My name is Rauno Aaltonen.' He must have been in his sixties.

We first showed [a working prototype] in 1997, at the Frankfurt Motor Show, in an event held offsite. It ran down a strip, with journalists not knowing what was going to come out of the box. The problem was, when it actually came out it was just supposed to drive by, stop for a moment and then take off, but it was surrounded by journalists getting up out of their seats and it had to make a hasty exit because people were trying to open the car.

It took a while to snowball, but as soon as the car started to look like it could really come out like that, and we had our first market-research clinics, everybody got really excited and said, 'We're going to come out with something that's really going to hit big time.'

Paul Chantry
There was always a feeling in Oxford that you needed two models to be secure, so if one wasn't selling so well you always had the other one. We'd always sold people that idea, and now the future of the plant rested on one model.

Gert Hildebrand
It's not a retro car. It's a clear, valid successor of the classic Mini, with the only exception that its predecessor was not changed for forty years. On other cars they change it every seven years. It would be very arrogant to say that if Issigonis would have lived in 2000, he would have come to a similar interpretation, because times have changed, regulations have changed, the expectations of what people consider luxury has changed. But we cannot ask Issigonis.

7 'He adored those lambs,' said Ronald 'Steady' Barker

Alec Issigonis (in 1986, aged eighty, with more than five million Minis sold)
And now this terrible thing they're using I think is absolutely monstrous. You know, in cars: the telephones!

Lord Stokes
The right thing to say is that we had in Alec Issigonis an innovator, a very brilliant innovator, a man who could do lateral as well as vertical thinking with the motor car, and who conceived the brand new breed of car we see in Europe today. One must give him credit for that . . .

The Mini had done well during the 1960s, its growing popularity at home reflected in Europe, where it accounted for seventy per cent of all British car sales in the decade to 1969. (Demand was met partly by the export of the car in kit form, known as 'complete knock-down', with assembly plants in Belgium, Italy, South Africa and Spain.)

The model range expanded rapidly. There was the 1964 Mini-Moke (originally designed as an air-drop reconnaissance vehicle for the army, but widely adopted as a beach buggy in Australia and occasionally as a babe-magnet in the King's Road); there were branded Minis, the Riley Elf and the Wolseley Hornet,

Opposite: Noel Edmonds celebrates the five millionth Mini in 1986

employing the same engines and chassis but with a new front and rear end and dashboard; there were the Countryman and Traveller with the bigger 998 cc engine and optional wood trim, and the Clubman and Clubman estate, introduced in 1969, offering concealed door hinges and wind-up windows as standard.

In 1968, there was a new force in British motoring: British Motor Holdings (the company that had been formed from the merger of BMC and Jaguar) merged with Leyland Motor Corporation to form British Leyland Motor Corporation (later BL). It was an attempt to counteract overseas competition, and the new company was led by Donald Stokes, who was more of a salesman than a man equipped to manage the second largest motor manufacturer outside the United States. It held forty per cent of the British car market, but its size soon proved a hindrance: its most popular cars – Austins, Morrises, Rovers, Jaguars, Triumphs and MGs – began competing against each other for the same customers, much to the delight of Ford.

When Donald Stokes took the helm in 1968, he had many action plans. One of the first things he did was to remove the Hydrolastic suspension from the Mini; it was just too expensive. And then he ordered the swift expansion of his design team.

Lord Stokes

. . . but at the same time I think you have to have a back-up team of developers who can develop all these ideas, and develop all these products down to the finest detail. In other words, refine them so that they get into production without any hiccups and difficulties. And I think the Maxi and even to a certain extent the Mini could have done with a little more refinement.

Issigonis had ideas . . . he liked the steering of the Mini which was rather like driving a bus, and that's the way he thought a car ought to be driven. If everybody else didn't like it, they were wrong. To have that mental courage is all right, but it doesn't always do in the marketplace when you're up against some very talented competitors. You can't have one man as a maestro running a huge engineering department of a huge motor manufacturer. It's got to be a team effort, and you've got to have specialists designing gearboxes and designing axles and engines, and then you've got to have an innovative engineer who can coordinate all that activity so that what comes out is one harmonious whole. I think the world had just gone on a few stages. He'd been the right man in his day, and I think events had almost overtaken him.

Issigonis was knighted in 1969, and he offered his retirement two years later on his sixty-fifth birthday.

Sir Alec Issigonis
I went into his office and I said, 'Lord Stokes, I'm about to retire.'

He said, 'You're doing nothing of the bloody well sort. You're going to stay here as a consultant.'

From then on things began to go from bad to worse and politics came into it, but being an engineer I'm not interested in politics.

When he did retire he asked for the flagship Meccano outfit, the No. 10, and all his future designs would take the form of

Sir Alec Issigonis at his retirement party in 1971 – with the first Mini
and other landmarks of his career . . .

*solitary fantasies. His mother died a year later in 1972, her last
days spent in a Nuffield nursing home in a Birmingham suburb.
Without her, his eccentricities deepened: one biographer
describes an instruction to any guests who called for supper not
to stack the dining plates, as these would then have to be
washed underneath as well. As the motor industry entered a
period of crisis in the mid-seventies, Issigonis became increasingly obstinate. His design for a Mini replacement, the larger
and less austere 9X, had not proved popular at British Leyland
despite the innovation of a 'gearless' engine and his reintroduc-
tion of sliding windows. Fiestas, Renaults and Golfs were filling the streets, but Issigonis refused to have any visitors park
within view of his house unless they drove cars he had designed
himself. In a pleasant irony, the office he was allocated at
Longbridge during his consultancy had a consistently leaking
roof.*

Alex Moulton

My friendship with Issi continued for a long time, but it failed very abruptly in 1969 [over a question of loyalties; Issigonis disapproved of Moulton designing new suspensions for British Leyland coaches].

Dante Giacosa, the brilliant Fiat designer [of the Fiat 500], had written in his book about how impressed he'd been with the Mini, but if you mentioned the name Giacosa to Issigonis he wouldn't give any credit to anyone else. It was a great, great pity. He had been so puffed up by his own success and royalty.

. . . and twelve years earlier at the Mini launch

Lord Snowdon

I had the great pleasure of him staying with us in Venice once. He's not a great traveller, always gets his clothes wrong, always arrives in a three-piece tweed suit whether it's in Switzerland in the winter tobogganing or in Venice in the heat of the summer. I led him into the centre of St Mark's at midnight. He'd never been there before, and it is a breathtaking sight, to see St Mark's empty. I said, 'What do you think of it, Alec?' and he looked at me and said, 'My dear, exactly like the Burlington Arcade.'

Then he got terribly cross in Switzerland, in the same three-piece suit. We used to go tobogganing, which was highly illegal, from Davos down to Klosters, down the main road. Having had huge arguments with various people about the design of the Mini, he was livid, because we had to be towed up by some renowned German motor car, which he didn't like at all, and he was screaming abuse the whole time.

The Ironmonger, BBC Radio 4 documentary, 1986

At eighty, and in failing health, Alec Issigonis lives quietly in a bungalow in a Birmingham suburb. He's never married. He's made no major contribution to any car project since his retirement.

Sir Alec Issigonis

What I see before me is stagnation. Everything's been done. Nothing more to be done. We've tried steam, we've tried electricity – well I have. I can see only stagnation. They'll remain as they are. You can't tell one car from another today unless you look at the badge. The only exception to that is the Mini.

Sir Alec Issigonis died on 2 October 1988, aged eighty-one, seven months short of the Mini's thirtieth birthday.

What sort of industry did he leave behind? An industry in turmoil. British Leyland Motor Corporation was effectively nationalised in 1975, following a disastrous performance in the preceding years. BLMC production declined from 1.7 million vehicles in 1973 to 1,260,000 in 1975, the company reporting a loss of £76 million. Now it was not just the Mini that was making a loss: every car sold was costing the company between £90 and £160. Increased competition and adverse sterling exchange rates meant that in 1975 one third of all new registrations in the UK were imported, overtaking BLMC's share for the first time. The Ryder Report recommended a government bailout of £1,264 million as the only way to offset the possibility of imminent collapse and its effect on the balance of trade and about a million jobs. Sir Donald Stokes was replaced as chairman and managing director by Sir Michael Edwardes, who came from running the Chloride battery company, and whose most notable initial task was finding a way to sack union leader Derek 'Red Robbo' Robinson. The appearance of the Metro in 1980 was a success, selling more than two million in its various guises (Austin Mini Metro, MG Metro, Rover 100).

And the Mini gallantly chugged on, offering new names and engines and marketing dreams to an increasingly uninterested public: the City E, the Mayfair, the Ritz, the Chelsea, the Piccadilly, the Park Lane, the Advantage, the Designer, the Checkmate, the Neon, the Sprite, the Cabriolet and the many new Coopers. Every big birthday saw a new special edition for the UK, and it said something about the car that the Mini 25 was made in an edition of five thousand, the Mini 30 in an edition of three thousand and the Mini 35 in an edition of one thousand. By the time they made the Mini 40 in 1999, the number was down to 250.

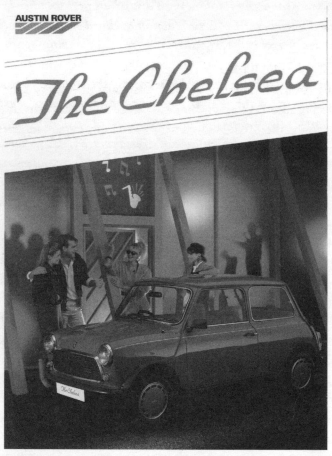

The Mayfair, the Ritz, the Piccadilly, the Park Lane and the Chelsea . . . all very, very similar

Alex Moulton

Issigonis was very hard up at the end, very sad. He had a good pension, but I think he had to sell his Edgbaston house to pay for the nursing home. You should talk to Steady Barker about him – I think he saw him at the end.

Ronald 'Steady' Barker

I got to know Issigonis very well personally. I felt very sympathetic because I'd known him ever since I did that trip in the Mini round the Med. He was the most entertaining, wonderful party man, he had the timing over dinner of a professional comedian. As a raconteur he was absolutely amazing. But he did like to be surrounded by yes-men. Moulton probably told you that in his later years Issigonis would have nothing to do with him, wouldn't even talk to him on the phone.

Steady Barker is eighty-eight, on the phone from his home in Purton, Wiltshire. We speak in the early evening, and he says he has had a whisky and is keen to have his dinner, but when he gets going on Issigonis and the old days it is a long and effortless flow. He says he got his nickname from a friend who was keen on postwar radio actor Eric 'Heart-throb' Barker and his catchphrase when romantically aroused: 'Steady, Barker!'

To be very frank, everybody thinks of him as being a homo. He probably was, but my own estimate of him is that he was probably a non-practising homo. He got on terribly well with Princess Margaret, because he was also a snob.

I think his mother was a millstone around his neck, but he stuck with her until she died. I probably still have tapes if I could only find them, and in the middle of the tape you hear her saying, 'Aw, but Alec, I . . .' in her Bavarian accent.

And he would say, 'Oh shut up, Mother, you know nothing about this.' I think she enormously inhibited his private life. She didn't die until he had become greatly restricted by what was believed to be Ménière's disease, it affects your balance. He couldn't go in a pressurised aircraft. I think the hearing from one ear was surgically removed from him. Then he told me later it wasn't anything particular to do with ears, but low blood pressure.

So he lived alone in this bungalow in Edgbaston, where he taught me to do water divining with a steel coathanger – but he knew where the water was, you see. Very, very entertaining company. Then he needed full-time help, so he was removed to a nursing home and he was room-ridden, if not entirely bedridden, and all his journalists and media people completely abandoned him once he'd become sidelined by the horrid British Leyland, that awful bloody mess at the top, these terrible people who took it over, Stokes, people like that. They started making ordinary cars like the Marina.

At his nursing home he spent his time doing crosswords in those cheap multiple books. In earlier times, on Sundays, he and Daniel Richmond's wife Bunty [of Downton Motors, the Mini engine tuning firm], they used to do a crossword on the telephone on Sunday mornings. Issigonis only did the simple ones if he got the *Telegraph*, not the main one. In his last months this was his main occupation, doing these effing crosswords.

I used to go up there on my Yamaha motorbike to spend an hour or two with him, and then when I got home he used to ring up, and say 'Steady, Alec here,' and by this time he'd be well up on the gin and Martini. We invented all sorts of names. When he was in an artistic mood he would be Annigonis, and when he'd had a bit too much it was Pissigonis. And when he had a pretty good appetite it would be Porkergonis. He rang up one day and said, 'Did I tell you that I hated lamb chops?' When he was a boy in Turkey they used to give the kids a lamb to love like a pet dog, and then after a few months they'd kill it and eat it. He adored those lambs.

Another day he rang up when I'd hardly got home and said, 'Alec here. Do you realise that one of the great events in the world has just taken place?'

I said, 'Well I'm sure there are quite a few – which particular one?'

He said, 'Germany abandons the swing-axle!'

Alex Moulton
Oh yes, I went to his funeral.

I think I had a Silver Shadow then, and out of respect to him I drove up to Edgbaston in that. Peter Ustinov spoke.

Peter Ustinov, at the funeral
His eyes, of a surprisingly intense deep blue, were recalled in the wide-eyed innocence of the Mini's headlights, childish but hugely sophisticated. The Mini was not only a triumph of engineering but an enduring personality, as was Sir Alec with his exquisitely caustic tongue and infectious merriment.

Ronald 'Steady' Barker

Alec died two weeks after I'd last seen him. Nobody else went up to sit with him and spend time with him. But I didn't get to his funeral because I was abroad, and one of my colleagues, another motoring journalist, said, 'I was awfully surprised not to see you at Alec Issigonis's funeral.'

I said, 'Oh yes, I couldn't get there because I was only coming back from abroad that day.' Then I said, 'By the way, when did you last see him?'

'Oh, it must have been two years ago.'

And I said, 'Well I saw him in the last two weeks, which is rather more important than going to his funeral.'

The last time I saw him he was out of bed, sitting in a chair in this rather impersonal, strange place, all on his own and doing a crossword.

Outside his home, Alex Moulton shows his visitor the best way for a tall man to get into a Mini – bottom first. The Mini is blue

with a white roof, and it was one of the last to come off the line in 2000. It was top of the line: wooden dashboard, cream leather seats with piping. It is parked next to a Rover 100 and a Toyota Prius, which Moulton cannot praise highly enough.

We drive off in the Mini through the country lanes. He explains that he has no children, and will be leaving his Hall to an engineering foundation to ensure continuity of his work. The car he is driving has rubber cone suspension, unlike the one he bought in the mid-sixties, which he has in his garage and which is still running well after more than a hundred thousand miles. He passes a new MINI, the BMW version from 2001. 'A perfectly good car,' Moulton says. 'But it's not a Mini at all.'

8 'The German target management system was something of a challenge,' said Donna Green

Elaine Butler (PA to the director of assembly and a former assembly-line worker)
So in 2000 they took out everything to do with Rover 75 and shipped it to Longbridge, even the desks out of the offices. It was nothing except JCB diggers. Then we had a lot of German guys come over, I mean loads.

The people that I worked with, it was the same as when the original Mini was here: they either loved the new car or they hated it, there was nothing in between. But I would have died for one.

Paul Chantry
The brief was to move Rover 75 without any interruption to the market availability of the car, and the same was true of MINI: to move MINI and not delay the launch. We agreed that we would build a stock of Rover body shells that would be sent to Longbridge, and we managed to build about three thousand and find buildings to put them in.

Then we didn't have a product for the best part of a year. We stopped in the summer of 2000, and we weren't going to be in full production of the MINI until the following spring, so some terms were thought up for people in the workforce who wanted to leave, and a lot of people went. We were left maybe with

Opposite: the first customer car on the production line, 2001

enough people to run one shift. The people who stayed were offered a deal: they were offered nine weeks off, and they would have to later recover those hours back by working more over the next couple of years.

So we were basically moving two factories at once. I think we had about fifty lorries going back and forth on a daily basis.

Donna Green (quality specialist)
It was quite a difficult time. There was a project team created called NOW – the New Oxford Way. NOW was a big change, designed to sweep away the old: this is how it's going to be from now on. But implementing the German target management system for people who weren't used to being measured was something of a challenge.

Many of the workers here thought it was a complete waste of time. 'Why do we spend all our time filling in charts when we could be working?'

Peter Crook (director of the paintshop)
In 2000, when we were waiting to put the MINI facilities in, Herbert Diess, the plant director, called me into his office and said, 'Do you have any idea what we can do with 250 or three hundred people for six months before the production starts?' I ended up with nearly six hundred people, and I set up my own business within the organisation, refurbishing the plant. We got all these people painting, doing carpentry, building brick walls, all these people with their own expertise, half of them DIY experts. We got them gardening, rubbing down the rails. We put some pride in them, so they felt they were creating their own environment.

The new MINI begins at Cowley in 2000

Paul Chantry

We weren't certain about the shape of it – will people like it? The impressions we got through BMW, through the project team, was that no one was really sure: it was always said, 'Well, we know it's new, and it's an exciting car, so we know sales will do this [makes an upward curve on a graph with his flat hand], but after that we don't know if it's going to do this [carries on going with the upward climb] or this [flatlines] or this [the downward curve].' So we were asked to put a flexible system in place so that if it took off we could keep up with it, but if it didn't take off at least we hadn't put in enormous fixed costs by hiring lots and lots of people.

The enthusiasm for the car grew, but it really only grew with its success. I'm convinced that the workforce weren't convinced

159

by it, and they weren't sure of BMW, almost whether to trust them because of what had happened to Rover. Or maybe it was suspicion rather than a lack of trust. There's a general perception that BMW hadn't done a very good job with Rover. With the support of Honda, Rover was slowly climbing its way back to break-even and being sustainable. BMW came in, didn't understand any of that, and then Honda disappeared and expected Rover to stand on its own two feet, and it collapsed. So there was that sort of feeling that if MINI was not an immediate success BMW would just walk away.

Peter Crook

Personally I was thrilled the MINI was coming back. I loved the Rover 75, but for me the Mini has always been Oxford anyway. The plant had worked really hard to be competitive and lean in its structure and operations. We had moved a lot further forward at Oxford than at any of the other Rover plants. The union and the management were far more of a team, and we had a much better understanding.

As a young man the Mini was your life. My first car was a Mini, and my first brand-new car was a Mini, and to have the MINI come back to where it was born was a dream.

Paul Chantry

The original launch date was put back a few months: the project team had taken it back because they had decided they wanted to do some re-engineering. So we got on with the installation. About three or four managers came with the project down from Longbridge, which was absolutely essential. We got the layout sorted, then we had some basic building work to fit it, and then you have to start moving the stuff down. Normally you would get about a year for that work, but we had three months. The Rover 75 building was new, and we didn't have to do much to it, basically clean a few bolts up. But the building next to it, which we needed, was an old building, and had been used originally for Rolls-Royce body production. That was in a state, and the first thing we had to do was to get asbestos out of the roof. We had to re-floor in the area where the robots were going to go, because the floor couldn't take the strain. And we had to put a new roof on because it was leaking – not much of a building, and all the underside of new MINI was going to be produced in there.

But our major concern was not getting into production, it was getting to the appropriate quality level – that's where BMW has very high standards.

Suddenly the place was full of Germans for the launch. All the managers who reported to me were German. My English managers had either left, or left because I'd asked them to. If I'd had bad experiences with them on Rover 75 and there was an opportunity to say, 'I don't need you on MINI,' then I took that opportunity. I would have discussions with my German mentor about this and he would ask, 'What do you think – you think this guy could do it?'

'Well, no.'

'So why kid yourself Paul? You can't compromise, the MINI has to work.'

I thought I was pretty good – I'd worked with Honda, I'd worked with the Germans, but in some cases the Germans even surprised me with the thoroughness of their approach. Making the front end of the MINI and getting everything to match is very complicated. Through the process, some things change. Maybe the customer wouldn't be too keen to know this, but you've got these two longitudinals that stick out the front of the car, and actually their dimensions change when you put the weight of the engine in. You have to offset for it. And then we had this wraparound bonnet, the biggest bonnet in Europe at the time, which had to fit everywhere. And that moved as well, so you had all these dynamics and variables going on that needed a lot of analysis. The standards which the BMW audit applied were tighter than any we'd ever known before, and were totally uncompromising. And just to be sure that they were totally uncompromising, guys came from Munich every two months to check. The complexity that my German R&D colleague said we should go through to understand these problems, I have to say, fazed me. It was tough – the guy was always up there in the drawing office putting in an enormous amount of effort, and he never went home at night.

Pat Nolan

It took a while to bring Oxford into the BMW world. It took us a couple of years. BMW are an engineering-driven company, and there was a lot of respect for our engineers. It was refreshing that you could speak to very senior people in BMW, whereas in the Rover hierarchy it was sometimes difficult to get above a certain level. With BMW, people always said we should call them Dr this and that, but they were actually very happy on first-name terms.

We obviously had to look at the layouts – the shop at Oxford was a different shape to the shop at Longbridge. Also, Longbridge was a massive site, and the investment that would

The Royal Mail celebrates a design classic in 2009

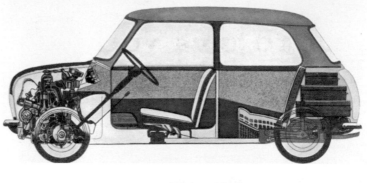

The first Mini, and what it took to make it

The electric MINI in Los Angeles in 2009, and a MINI concept for another age

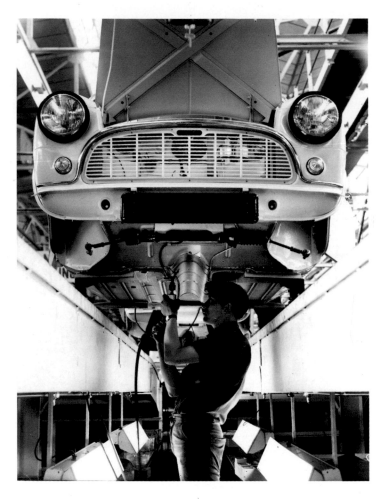

Making the Mini old and new: a worker drills the undercarriage . . .

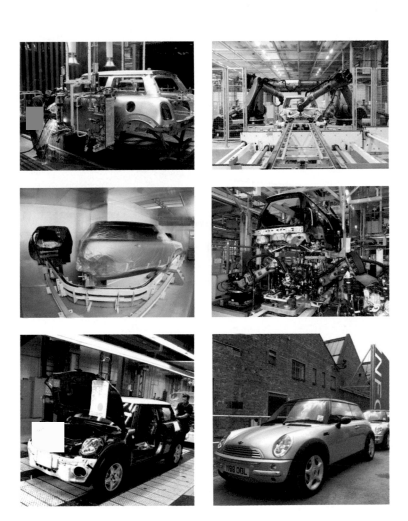

. . . the robots take over in the paint shop and the assembly line

The little dog laughed to see how easily she parked

Mum's in charge on this trip! And the Mini-Minor is such a dream to handle,
she can nip in where the grocer's boy usually parks his bicycle! It cuts shopping
costs too—why, you could carry enough stores for a *month*!
There never was a small car which seemed so big on long journeys—yet
remained so wonderfully neat and parkable in town traffic.

Breakfast will be served on the Mini-Minor

No time to close the garden gate! But never mind—this is a best-foot-forward
kind of car that weaves its way through dense town traffic. She'll be hearing
the school bell before she's even finished her toast!

"As a matter of fact, this *isn't* my favourite car of all time."

"This is one of the latest Minis.®

"My all-time favourite was a Mini I bought a few years back. It was tremendous fun. So much so, I got another.

"But I thought you could never recapture the thrill of your first Mini. Until I saw this one.

"They've put in new wall-to-wall carpets, soundproofing, new seats and controls, a new, smooth suspension and they've given it Supercover protection.

"My favourite car of all time will always be my first Mini.

"If your next Mini's your first, you'll soon see what I mean!".

Twiggy

Welcome back to a better Mini.

 Mini

From Leyland Cars. With Supercover.
' Mini' is a Registered Trade Mark

Selling the car the old-fashioned way . . .

MAN WANTS TO HAVE A THOROUGHLY GOOD POKE AROUND THE NEW MINI.

GOES TO DEALERSHIP ON 7TH JULY.

STARTS POKING.

THE END.

On Saturday 7th July the adventure begins when the new MINI is launched in its own showrooms alongside BMW dealerships throughout the country. To find your nearest dealer and have your own poke around, call 08000 836 464 or visit mini.co.uk

IT'S A MINI ADVENTURE
From £10,300 on the road.
Car shown MINI Cooper £12,180.

. . . while the new MINI Adventure takes off at dealers and the London Roundhouse

The dashboard gets an update . . .

The luggage performs its unlikely tricks

Old and new commemorate the triumphs at Monte Carlo

have been required to keep that going . . . whereas Cowley was quite small. But the buildings were quite old, and BMW brought them all up to their standards, so if you go into our bodyshop now, it's like being in a BMW bodyshop. We're not a forgotten leg, we are a BMW plant.

Paul Chantry

In the 'dwell period', before we actually got into production, the time came to renew our workwear. In Oxford it was always grey with the Rover badge on it, and in BMW it had always been grey or blue overalls. The discussion got to the level of the directors talking to the PJC [the Plant Joint Council, the senior trade union representatives]. We got a strong feeling from the

New uniforms, new car: the first body shell in 2001

PJC that they wanted to look like BMW workers, that they wanted to instil confidence in the future of the plant, and there was a strong argument from the head of the PJC about why people expected to be walking to work wearing these grey or blue overalls with BMW on them.

But Herbert Diess, who had been the boss for about six months, had a totally different view. He said, 'We're developing the MINI brand here. You guys may not see it yet, but MINI is what this is all about. Being part of BMW is nice for us, but what we are here to do is to sell the brand MINI.' It always struck me as an issue of trust.

So he got some stylist in Munich to come up with some alternatives. You know the corporate identity of MINI launched in the showrooms? Lots of black and then strong primary colours, the oranges and yellows and greens. He got some workwear made up in these colours. We had a little mannequin display of these things, and I remember the head of the PJC saying, 'People walking to work in that will look like clowns. They'll be embarrassed to wear it.'

But Herbert loved yellow. He had a horrible liking for the colour – he used to have a yellow motorbike and walked round the plant in a yellow jacket. In the end we had a vote, and it was really close, something like six to five. So by a majority of one we voted for the funky workwear. The amusing thing is, having had all these discussions about this clownwear, within nine months, when the car was launched and it was a success and you could see the continuity between the branding of the plant and the branding of MINI, we were walking round pleased as punch, talking about it being funky workwear for a sexy funky product.

9 'They're made of silver foil and they're twisted,' said the promotional material

Ian Cummings

From the road you think that people just screw cars together and weld them and paint them, but there's a whole world going on in here.

Ian Cummings, responsible for associate training, was seven when the Mini was born. He is sitting in his office above the final assembly line. Plant Oxford has been his workplace for almost forty years, a statement that should not come as a surprise.

He is the son of Jean and Eddie Cummings, the chief safety officer and Rexine trim-shop sewer. Before them, his grandfather was here as well. But Ian never thought he'd end up on the plant too. He was a bright boy, attended the same Oxford High School for Boys that once educated Lawrence of Arabia, and just when his parents thought that everything was going well, he went off the rails, 'like young men do'.

I was sixteenish. I was much more interested in going into Oxford for discos. I made a bit of a balls-up of my education, and anyone in that sort of environment, when the dad says, 'I've had enough, get out to work,' most people ended up in the car factories at Oxford.

There were twenty-six or twenty-seven thousand people here, producing perhaps three or four hundred thousand cars a year. I joined in the summer of 1969. I was taken on a tour, down to the building where they currently build the MINI. There was this huge great machine shop and press shop there. At one time it was the biggest press shop in Europe. A huge lot of activity, it was all tarred wood-block floors. We were building Rolls-Royce bodies, Hillman Minxes, Triumph Spitfires, MGBs, Morrises, Rovers.

Things go into the back of your head: 'This all looks so old-fashioned.' You wonder, 'Are they reinvesting correctly here?' It was like I expected to see my grandad working in there. These massive flywheel presses going round, still churning away and thundering up and down and shaking the floor. I've got photographs from then, but that was in the 1930s and I thought that it ain't much changed.

Paul Chantry
I came here for an interview on my twenty-third birthday. From where I'd come from, it was all a little bit antique. My first boss described the Cowley plant as the world's largest village: grandees, the senior managers who had been here forever, a hell of a lot built on personal relationships, lots and lots of family tradition. Lots of big middle-aged men who ran production, and they walked around in their white coats with different coloured collars – green for technical, red for production, blue for maintenance.

The general impression locally was that Cowley wasn't terribly welcoming to graduates. I think Lord Nuffield and the people who managed the company were probably more into homegrown manual talent. That's the way the company developed: they started off as garage mechanics and turned into automotive engineers.

My first job was in Pressed Steel Fisher. The first time I went in, they were making Maxi body shells. The Marina was the next to be launched. It was like walking into Dante's *Inferno*. It was all manual, and my first impression was of a noisy, dirty, smelly place. And then you went into the press shop, and the noise was so tremendous. The only way to be heard was to grab hold of somebody and shout in their ear.

And the other thing of course was the bicycles – all the people streaming in and streaming home through the gates. There was no flyover then, so by the roundabout it was just a wave of bicycles.

Generally you would have a day shift and a night shift. The Maxi hadn't sold anything like the expected volume, so I guess they weren't working anything like the normal two-shift pattern, but if you went up to the Rover 2000, which was at the top end of the site in a building that's now gone, that was absolutely buzzing. There was this different village that came on at night.

It was all part of British Leyland, but it was all run as different companies. They started to come up with the idea of 'the division'. In my first seven years here, we had something like seven or eight company names. I often joked that I moved my desk five yards, and had seven different company titles.

Making them the old-fashioned way: Issigonis clambers in again for the 2 millionth Mini

'The 2 Millionth Mini: A Giant Promotion for our Small Success', Longbridge dealership promotion, May 1969

After ten enormously successful years for the Mini, we've reached the two million mark. It's a big occasion for all of us. And we most certainly couldn't have done it without your continued sales support. But now's no time to rest on our laurels. This booklet shows you the full promotional kit we've designed for the event to help you make a big sales impact right on your own showroom doorstep.

A Foil Twister. You've probably seen these before. They're made of silver foil and they're twisted. You can use them in either of two ways. One, you simply hang them in your showroom. Or two, we'll supply you with a small motor that rotates each twister to give it an eye-catching 'corkscrew' effect. Without the motor, the cost to you is 14s 6d per twister. With the motor, £2. 0s 0d.

B Giant Numerals: '2,000,000'. These letters are the same height as the wheels of the Mini. Incidentally, the pieces are strutted at the back. You can range these numerals around the front of your main showroom model or you might want to use it in the window. It's up to you.

C Pelmets: 'You get 2,000,000 friends with a Mini'. These are really window stickers. But they're big ones – two pieces each 5 feet long by 15 inches wide.

D Car Headboard: 'The smallest success in motoring history'. Just set it across the top of the car or place it sideways. You've seen all we're doing for you. Now show us what you can do for the Mini.

Paul Chantry

The thing that absolutely shook me about the Minis was that when they put them together they didn't fit.

The first time I saw Mini production was in 1970 at Castle Bromwich, who were making the trim body shells for the very high production at Longbridge. Castle Bromwich is now Jaguar.

I had several jobs here in quality control. One was planning for the dimensional accuracy of the body for new models. I began on the Allegro, then the Princess, and then later Metro.

On the Mini, if you recall the original boot – it was just a hole in a single pressing, which in my eyes would be a simple thing to control. Once you got the tooling right, it would all fit and the trunk lids would look perfect. But they used to put these trunk lids in with what they call a stretcher bar. This was a mechanism, a parallelogram of wood or metal and a bar, and you'd put it in and stretch it. The same would apply to the doors. They put the doors in and they didn't fit, so they'd get these bars on them and bend everything.

I went in to see the guy who was responsible for dimensional control in the plant, and he said, 'This car has been in production here for ten years, and there's not a single thing that's right to drawing. And I'm now going to put it all right.' But of course he never did. He put a lot of effort in to overcome some of the basic problems.

Part of it was bad technology, and part of it was, 'It's good enough – the customer won't notice it at sixty miles per hour.' It was that sort of approach to car manufacture – a bit of bodging.

Letter from Cedric Scroggs, 28 April 1977
Mini Training Programme

To all Austin Morris Distributors, Main Dealers and Retail
Dealers

Dear Sir,

Further to my previous letter of 28 March, ref CAS/
PBS.PT1, please find enclosed the second part of the Mini
Training Programme.

Yours sincerely,

CA Scroggs

Marketing Director

Enclosure: Mini Seminar Leader's Guide
The widespread demand for more economical motoring and
the recent introduction of the Ford Fiesta have led to an
upsurge of interest in the small-car sector of the UK market.
The purpose of the 'Mini Training Programme' is to help you to
better equip yourself to sell the Mini in the increasingly competitive prevailing conditions.

On the following pages there are seven major areas for con-
sideration. In each area the syndicate should do the following.

1   Identify the advantages of the Clubman Saloon over the
    Fiesta 950L and the Fiesta's advantages over the Clubman
    Saloon.
2   A summary should be made of how best to present that
    aspect to the customer. The summary should concentrate
    on Mini's strengths, but also include your ideas for over-
    coming known criticisms and weak points.

50 per cent of all small cars are owned by two-car families and 50 per cent of all small-car owners are women. As a first step in exploiting this source, the syndicate should establish what aspects make Mini a) a good second car and b) an easy car for a woman to drive. In discussion with the other syndicates you will arrive at a plan of action for increasing Mini sales to the woman drivers of two-car families.

Paul Chantry

Once we had a Mini sent back to Cowley by a customer. If you remember on the original Mini the seats just tipped forwards, but on this car the seats wouldn't tilt forward on one side because they fouled against the pillar behind the door. We had this car parked outside our office, and the quality manager arrives, and I give him the keys so he can do his analysis and we can decide what to say to the customer. I watched him from the window, and he had a tape measure. He put the tape measure across the body, from one side to the other, and then came up and said, 'Well, there's nothing wrong with the body, it's perfectly all right. There must be something wrong with the seat.' That was the level of analysis. There's two things wrong with that: a) it's not analysis, and b) the concept that you can measure a body with sufficient accuracy with a tape measure . . .

To me that was a bit of a shock. I'd worked in aircraft, where the tolerances were very tight and the testing regimes were absolutely phenomenal. Here it was not professional and it was not done on a scientific basis.

Opposite: Making it fit, come what may – Cowley production line

Ian Cummings

I was bright enough to talk my way into what was a commercial sort of apprenticeship, in Pressed Steel Fisher, the plant we're in now. The motor industry gave fantastic careers to people with all levels of education, and although I didn't actually realise it at the time, I had ended up in a place that offered me huge opportunities. During my apprenticeship I actually went into the Pressed Steel Fisher form design department. The bureaucracy in these places was enormous, and you had to have standardisation to do all sorts of things, and there was actually a department with guys sat at drawing boards designing forms.

Paul Chantry

There were quite a few attempts to improve things. At some point we had a new director of quality arrive, called Brigadier Charles Maple, who came from the Ministry of Defence. It was around the time that Michael Edwardes was in charge. Brigadier Charles Maple suggested that we should apply the same quality standards that had applied to defence contractors. You had the dedicated few who were trying to implement this, who could see the sense of it, but to say it had any impact on the business would be a total exaggeration.

There was quite a lot done on trying to spend money on the right things. The machinery was so out-of-date, the technology was so poor, so quite a lot was done through the seventies. With the launch of Metro [in 1980] there was a lot of money spent on automation, and new machines that made the body tooling and measuring machines. A lot of money was spent – I

Mini model meets mini model for the 4 millionth car

guess it was the government's money, the Ryder Report's money.

I think things did get better. But even then the Metro still had the old A Series engine in it [one- or 1.3-litre], so there was a limit to what you could do, and the cars were still realistically no match for something like the Ford Fiesta.

Roy Davies
The problem that all models suffered from is that prototype vehicles are not usually quite like the production ones. It might be better now with the way they do things but it was totally usual to get totally caught . . . I mean the Maxi was a disaster,

what with the change from prototype to the production. The Princess – exactly the same. It had been through every known test. There was this thing called a pothole test where you drove a car through the equivalent of the deepest pothole drain you could find and it had to withstand twelve smashes flat-out without breaking up – which it did. But when we came to production cars, the big dash panel looked just like a lace curtain after about ten drives through this thing because there had been changes made from the prototype. The difficulty you always have is, you can make any shape in the world by hand. What you cannot do is make that shape always through a press, and so subtle changes within the press can make all sorts of differences.

Paul Chantry
With the Mini things didn't improve a lot either. The manufacturing process didn't change, the car didn't change, and as employees we never understood why it was never replaced with another car. It just carried on in its own sweet way. In the late eighties the idea was, 'Well, we'll just keep the car running until we can't.'

Pat Nolan
In 1988 I went to Longbridge for twelve years. I worked on the old Mini a little bit, we did the Mini Open, where we cut the roof off and put on a cloth cover, like a big sunroof. At Longbridge they cherished the Mini. They knew it would be phased out at some time, but until BMW came along and started developing the R50, the new MINI, I don't think they ever realised it definitely would.

The first Mini and the thirtieth anniversary special edition (top); the forty-year press campaign with Union Jack roofs

Roy Davies

We couldn't get any money for doing it up, so it continued on and on and on. As Harold Musgrove, the director at one time, said: 'It will die when the customer is ready for it to.'

Pat Nolan

If you didn't deal with Longbridge – and we were lucky enough to – the thought process from Cowley was 'They're nothing to do with us, and they're not as good as we are down here, and we build the quality and they build the volume.' But when you go up there and deal with them, it was a good place. I do think they realised that there was a lot more technical know-how at Cowley, but there was a buzz about Longbridge, they built a lot of cars, and they were committed to what they were doing. It was no different really to down here. They had their union problems, and strikes every other week almost, or that's what it seemed like when you were younger. When my manager said, 'Ee want you to go to Longbridge,' I said, 'I don't really want to go.' Because I'd heard all these stories. But I went up there and it was the best move I ever made.

Paul Chantry

It was a question of doing whatever was necessary to meet regulations, emissions regulations or whatever. If it was too expensive then obviously it wouldn't be able to continue, but in the end it found a sales market in Japan. If you went to look at Mini production, to my mind it looked the way it had ten or twenty years before, and if you went into the bodyshop it was just old men with grey beards building the car the way it had always been built.

At the end it was selling in such small numbers. With the Mini it was really the brand that somehow survived.

Longbridge promotional booklet, 1996
Over the past 10 years, Longbridge, along with its sister Rover Group plants, has undergone a significant revolution in Total Quality. Following Rover's philosophy of 'Success Through People', the plant has achieved a reputation not only for the enhanced quality of its products, but also for the associated improvements in its overall quality of industrial life.

Having achieved the unusual feat of becoming a classic in its own production lifetime, the Mini continues to garner accolades. It was voted car of the century in the centenary celebrations of *Autocar* magazine, and then the Mini Cooper took a similar award from *Classic & Sports Car* magazine.

Annual production is now running at around 20,000 a year, with strong demand continuing particularly in Japan.

The last old Mini was produced at Longbridge on 4 October 2000. It was number 5,387,862, a red-and-silver Cooper Sports, which Lulu drove off the assembly line in dry ice and flashing lights to the accompaniment of the theme to The Italian Job. *Draughtsman Jack Daniels explained that he was both proud and sad, and that he had believed that his car would last ten years, twelve at the most; Tony Ball, the man who had launched the first car in similar circumstances forty-one years before, spoke of Wizardry on Wheels (the poodles he had squeezed into the Mini at its launch in August 1959 had, in the mythical retelling, become Afghans).*

Kevin Howe, chief executive of MG Rover Group, remarked on how everyone had tears in their eyes. 'But this is not just the end, but also the beginning.' He spoke of a bright new future, and of his excitement over the Rover 75 and a new MG range. But his optimism was misplaced.

An ending and a beginning, but which was which? The last old Mini at Longbridge with the Rover 75

10 'Then you had the things about people caught sleeping,' said Bernard Moss

Chris Bond

We got union recognition here in 1934, after many miners had walked up looking for work on the plant during the Depression. They were living in tents in the surrounding fields, and it was mostly the women who organised it.

Chris Bond, fifty-five, deputy convenor at the MINI plant, started working at Oxford in 1972 with the introduction of the Maxi and the Princess. He is reliving it all in a conference room near the corporate publicity and marketing office, recollections that include owning a basic Mini 850 when he was eighteen and graduating to a more powerful Cooper. He became a shop steward in the late 1970s, carrying on a family tradition that began in the Welsh mines.

I was born in Aberfan in Wales, the same place the disaster happened in 1966, and my aunt's house was destroyed and my mother knew people who died. My mother came from a family of fourteen. The boys all went down the pits, but they sent a coach down for the young girls when they left school at about fourteen, and brought them up to England to work in the big houses so they could send money home.

My grandfather was the leader of the lodge in Merthyr and a staunch communist. During the 1984 miners' strike, Maerdy was unofficially twinned with Oxford, centred here at the Cowley plant, and we did collections for the children and sent food parcels.

The women at the Cowley trim shop celebrate Easter, 1976

I was brought up in Reading, and my first job was working in a boutique selling clothes. I was a mod then and you wanted to look good. That was £7 a week. But then I met a girl who was at the University at Oxford, and that's how I came to move up here. I went down to the local labour exchange, and they were hiring. I didn't know what Pressed Steel was, or about the car factory, but I got the job and I've been here ever since. Before I came to work here I was earning £22 a week, and here I started on £49 a week, and within a fortnight I was on a permanent wage with an extra £20. So it was £69 a week, and I couldn't spend it. I was paying £8 for a room in Headington – can you imagine, twenty-one years old with that sort of money in your pocket?

By talking to people I began to understand what trade unionism really was. I began to do a home correspondence course. I've always been interested in political history, and I've never been frightened to speak up for people who can't speak up for themselves. I remember my mother telling me stories of how she was treated, working for pennies, getting up early just to light fires so the gentry could get warm.

Ian Cummings

In those days, the trade unions and the workforce always thought the management was trying to get one over on them. They were always resistant to change, because they thought the management was trying to get more out of them for nothing – just a non-meeting of minds.

Our manufacturing management teams – highly responsible, highly paid, stressed-out people – they spent far too much time not thinking about the product but sitting in meetings having endless rows with trade-union officials about whether there were enough tea machines in place and 'Fred's been disciplined because he's absent, so we're all going to go on strike.' I think that partly destroyed the reputation of the cars we made.

Chris Bond

The thing I remember more than anything was the comradeship. On the first day I started here in 1972 the first person you see is your shop steward, and he took you under his wing and

Every spare part you could possibly need at the Cowley plant in the sixties

became a father figure. And if the managers tried to make you do something more than you should be doing, the shop steward would be there saying 'Don't do that, it's not on . . .' In Pressed Steel, the industrial action wasn't as bad as it was at the assembly plant in Morris's. When you speak to Bernard the convenor, he was at the Morris assembly plant and he'll tell you about the strikes there. You never worked a full week.

When I started I was finishing the fenders and the bonnets on the Maxi and the Princess. This is how things have changed: I was fitting a fender loosely on one side of the car, and there was someone fitting it loosely on the other side. And there would be two people loosely fitting the bonnet. Then four other people would set the fender and bonnet on both sides. Obviously that changed, and when I was working on the Montego I was fitting the lot on one side.

After the Maxi I went into the paintshop for a number of years, and when that modernised I went back into body-in-white to work on the MG, the Montego, the old Rover 800, the new Rover 600, fitting the rubber round the door, a mindless, terrible job, and physically hard. Then on to the Rover 75, and that's when I was elected area deputy and when the Germans came over. I had full-time facilities as a union man in an office.

When we built bad cars I don't think it was down to the workforce. On the Marina and the Maxi, people conscientiously did their work, and the reason the cars were so bad is because they were engineered bad. I remember on the old Rover 800, going over to the airfield at Oakley with glue to stick the trim back on. And then you'd be taking the mice nests out of the boot because the cars had been sat there for so long.

Ian Cummings
When I began in industrial relations here, my role in all of that was to ensure that we had properly run smooth meetings and that people discussed things sensibly. We tried to do the outrageous thing of introducing team leaders who weren't part of management. Team leaders were hourly paid guys off the shop floor – Christ almighty, that was unheard of – spies in the camp and all that nonsense. One of my jobs was trying to introduce that to vehicle assembly over the road. The sort of industrial-relations atmosphere we were working in meant that a guy wouldn't move

twenty yards from this track to that track to keep it running, but he'd be happy to stand idle. 'That's not my job over there, I'm not doing that, but I'm happy to stand here and do nothing, because they're not getting cars to me.' That lunatic attitude.

When you work in a factory that big, with tens of thousands of people, you seem to get institutionalised – particularly the workforce, but also a certain layer of management. 'This has always been here and it will always be here in the future.' They don't see the pounds, shillings and pence and the reality of life. They seem excluded from it, as if in a prison. 'We have all this silliness, we go home again and they pay us.' Very frustrating: these very bright capable people completely clouded to reality. And there was inertia. Like a great big supertanker, you couldn't change course.

During the Thatcher years the management became very aggressive and forceful and shouted at everyone. So immediately you start polarising people again: 'We have to do this!' and bang tables and be aggressive, and then, 'The workforce won't be spoken to like that!'

There were some manufacturing managers here that behaved in the most appalling way for educated people.

Chris Bond

A lot of the things we take for granted now – the canteen facilities, the sick pay, holidays – is down to the union organising on the plant. But Oxford has always been very political, and always very strong trade union-wise, helped by the student meetings in the town. We used to have big May Day rallies and the students

Opposite: The shop floor at Innocenti, where they assembled the Mini under licence

would be involved. But the left-wing factions all homed in to the car plant and infiltrated it, and that's why you had all the wildcat strikes, because everyone tried to get their point of view over, and everyone believed that next week was going to be the revolution. And in a lot of ways we believed that the working class did have that sort of power.

Bernard Moss (plant convenor)
When I started in the seventies the unions were all-powerful, there's no doubt about that. I don't know whether it was weak management or what. The powerful part of the plant in those days was E Block Paint, which was where they began with the Mini.

We had so many political groups with their newspapers on the gates. The Communist Party, the International Marxists, International Socialists, the Workers' Revolutionary Party, the Socialist Workers' Party. It used to be interesting at shop stewards' meetings because every group would have probably two or three shop stewards each, and they spent more time arguing amongst themselves than with the management.

Bernard Moss was born in Oxfordshire in 1951. After a stint in the blanket mills he entered the car business at the age of eighteen, making heaters for the Mini at a factory in Witney. He started at Cowley on the new Morris Marina, and swiftly became involved in the unions. As with Chris Bond, it was in his blood.

You've heard of the Tolpuddle Martyrs, 1830s, but there are books also referring to the 1870s and the Ascott Martyrs.

There was a strike on at the local farm, and the wives went out. The farmer was a magistrate, they were arrested, sixteen of them, one of them had a child. Most of them were under the name of Moss. They were taken to Chipping Norton police station and word got around, and the story goes that the police station at Chipping Norton was besieged by two thousand farm labourers. Eventually they were released and pardoned; Queen Victoria gave them a petticoat and a shilling each. They were marched back to the village of Ascott under Wychwood with a brass band, and they were heroes.

I suppose I don't like seeing injustice. Not everybody can speak up for themselves, and I was always the one in there if somebody was getting a bollocking and they weren't in the wrong.

In the seventies I can remember this article in the *Sun*. They were saying, 'All these strikes . . . giving the work to competitors.' The Japanese were starting to come into their own. We'd just had a dispute over water on the floor, and actually it was quite justified. There was an electric cable, and this water hadn't been cleaned up quickly enough. They did have a stoppage over it, one of those stoppages that got ridiculed. The *Sun* did a cartoon of these little Japanese people with these garden sprayers going round the outside of the plant, and these workers saying, 'My God, water! All out, lads!'

Then you had all the things about people caught sleeping. Without a doubt there were some absolutely stupid disputes, but my argument is this: in 2004 we celebrated seventy years of TU recognition on the Pressed Steel site, and when we did this presentation we had one guy talking about the conditions that people worked under – no health and safety, the only holidays you got were statutory holidays. The working day was any-thing: you'd come in and line up and they'd say, 'I want you, I don't want you, I'll have you . . .' On other days you were expected to do thirteen hours.

They had an open day up at this plant in 1986 and my dad came. He's dead now, but he was in his sixties then. When we went round the body-in-white building together, he said to me, 'Have they cleaned this up just because we're coming round then, boy?' I asked him what he meant, and he said, 'When we had slack days they used to get us with paint scrapers scraping the oil and that off the floor, it used to be about that thick . . .'

You've got moaners, you'll always have moaners, I think everybody wants to be a capitalist at heart, don't they, all bags of money and all of that, but if you look at the conditions that we've got on the site now, after all those years, they're probably the best in the car industry in this country. That's the achieve-ment they got after all those years, our forefathers, and some of it had to be done through strikes, unfortunately.

And then when we went into the eighties, it went the other way. They brought all these managers in from the Speke factory in Liverpool where they made the TR7, which had closed down. Some of them were out-and-out bastards, the way they treated people, and under their regime we had the longest strike I've ever been involved in. It went on for four weeks and a day and it was over something silly, but it was all the other things that led up to it.

But then things started to improve. The unions and the com-pany started to get together and say, 'If we want to survive, we've got to start doing the business better.' Up until 1986, when we were called to a meeting, the trade union side sat here, the management sat there, and the management would say, 'Right, we're going to bring in this new shift pattern . . .' and then the shop stewards told the members, who either accepted it or said, 'We ain't fucking doing that.' But in 1986 we decided to negotiate first, rather than afterwards.

Chris Bond

Of course you start to grow up and have a family and responsibilities, and bills have to be paid. It sounds a bit defeatist, but it's just the reality of growing up.

In the late eighties there were threats of closure. The convenor, a guy called Ivor Braggins, had to go to Japan to do a deal with Honda to get the Rover 600 on site. And that's when we started to calm down and people changed their attitude. The company was allowed to manage the business, and the trade unions had a more sensible approach.

Bernard Moss

We had so many bad experiences with Rover senior managers. It did get good towards the end, but you couldn't always trust them on their word. You'd get in a meeting and you'd negotiate something and as soon as you got out the door it's like, 'Well that's got rid of them, we'll do what we want now,' and I didn't like that, I'd sooner have a 'no'. And with the Rover production system – whatever volume of cars you did, it didn't actually mean you sold them. You could be doing two hundred thousand a year, you could only be selling 150,000, so you stacked the other fifty thousand in a field and hoped that somebody would come along and buy them later on. Fortunately, BMW doesn't work like that.

Chris Bond

It wasn't until BMW came on board that we actually developed a partnership. I'm not saying we're in their pocket or anything, but we have a better understanding.

We're given information far earlier, of quite a confidential nature, and we are part of the European Works Council, and we have a union guy who sits on the supervisory board and votes in German. It can be frustrating, because we know things that people on the shop floor don't know, and when rumours

The last thing we want to do is walk out: proud of the product in 2001

come round you just can't share that information with them, because they don't understand the bigger picture.

Bernard Moss

I've had people come up to me and say, 'I'm thinking of becoming a shop steward, how much extra do you get paid?' You don't get paid. In fact, a lot of the time it cost me money, because I was always having to go to meetings outside shift hours and running the car. I took over as plant convenor in 2003, and that's full-time with full-time facilities but I don't get extra money either.

I like doing the job, but there are times when I wonder if I'm getting too old. One of the problems I've got at the moment is that everyone thinks they can do the job but nobody wants it. We've got probably about forty or fifty shop stewards across the plant and there are some encouraging signs amongst one or two of them that need nurturing. But you don't have what you had in my day where you'd have two, three or sometimes four people battling it out to be shop steward. Often you get people who think they can do the job better, and we get a bit of stick at mass meetings on pay review: 'You bunch of bastards, you should have done this.' Some of it gets quite personal, but it's only half a dozen individuals, which you expect.

Chris Bond

One of the biggest changes under BMW was Working Time Account. They introduced it just when Longbridge was under threat. If they needed extra volume you would work the extra hours, bank them, and then take time off later. In Germany this idea had saved them in the eighties, and I think that's when BMW saw the trade unions as the saviour.

Oxford was against it totally. We were the only plant against it – everyone else was convinced that this was the only way forward.

On the whole it's worked out OK, but there is still hostility towards it. If people had the choice now they would do away with it. People like having the time off, but there is a little bit of abuse, where the company use it for the wrong reasons. You don't get paid overtime. The WTA is forty-five minutes extra on your shift, and that would go into my account. I can then add those hours up, and sometimes I can say, 'I want to have an extra week's holiday,' or, 'I want an extra month off because I'm building an extension.' It has worked well in those circumstances, people have enjoyed it, but it's been abused because the company has used it for breakdowns, and we have a separate agreement for that.

Paul Chantry

There was always a concern from the trade union side about the casualisation of labour. Their model was, you come to work, do regular hours, you know you're going to get your salary, and that's the end of it. Whereas the WTA system says you work the hours you're needed according to demand and the shift pattern, and if we want some more from time to time we'll ask you to work more hours, but at other times we'll lay you off and then you'll work fewer hours. If Herbert Diess had tried to bring that in fifteen or twenty years before, the work-force would have just laughed at him. It's only when you look back on it that you realise the vision. What he did to change the flexibility of that plant . . .

Bernard Moss, in September 2008, before the recession had an effect on the plant
Sometimes the more militant people don't like me saying this, but why not – you've got to have a company that's successful and we've got one. Yes, they're an employer at the end of the day, out to make money, but surely that's got to be good. You can argue about how much profit, but the plant's profitable, and in all my time at the factory I've never known such a phenomenal car. I mean normally you have three-ish good years then it starts to go down, you start getting rid of people, dropping lines. That's not happening, although we're not at full capacity. Full capacity at the plant is 260,000 cars. Last year I think the capacity was 240,000, 237,000.* We've always gone up each year, except one year when they had to take three weeks out to do a load of facility work to get ready for second generation.

I'm obviously privy to a lot of development, rebuildings, and stuff that in my position as the plant convenor is confidential. I can generalise about it and say to people yes, I know of plans

* The total production figure for 2007 was 237,700.

for 2020. I'll be long gone. People often say, 'What's this about shifts coming off?' because the weekend shift coming off was the big rumour for some time. I've got a son working in the paintshop and he said, 'Come on dad, there's got to be something . . .' I said, 'Look boy, there's nothing at the moment.' But you can't guarantee you can maintain that success for ever and a day. Everything seems to be governed by Munich. Munich is always there, governing all these car plants and whatever Munich says you've got to do it, and if you miss and don't quite make that target, 'Oh God, we've got to report to Munich now.' And I don't think we've ever missed a target.

Chris Bond
The other bone of contention was the agency labour they introduced. We induct the agency workers in the union, and we also conduct their pay and conditions agreements. We try and look after them, and we fight for contracts. But we've always had more than we'd like. The BMW philosophy is to protect the core workforce. One thing they will never admit to is redundancy. They hate the word redundancy. So they use agency labour, and if there's a fluctuation in volume, if there's a downturn, then they can get rid of people but their hands are clean.

The problem we've got is that everyone who comes here wants to work for BMW – they don't want to work for an agency. Because of the prestigious name and because on the whole it's a good company to work for. It isn't like the Rover days, where you'd come in and the first thing you'd do is look down the line to see where the nearest gap is to see when you'd have time to put the kettle on. They actually manage the company now – you come in, you work, and the only time you stop is when it's your official tea break.

We haven't had a dispute since 1994. It's come close a couple of times, and it can get a bit fraught, but we've always managed to negotiate. The last thing we want to do is walk out. We're here to get the workers good pay and conditions – we don't want them to lose money. I've learnt over the years that if you go out on strike it takes twice as long to get back what you've lost.

You have to be pretty foolish to lose your job. One of the biggest problems we've got is absenteeism. It's a little bit upsetting. We've got a first-class sick scheme. When people work here from the agency they're usually here two to three years before they get a contract. And they never have a day off, because they're all trying to be shining examples. We have a matrix system, which although not perfect is the fairest way of people getting jobs, as opposed to the company picking and choosing. As soon as they join BMW they have to do another twelve months' probation, and they don't take any time out. But as soon as they get the contract, they go sick, because we have a two-year sick scheme. There are genuine cases, if people fall sick or have an injury, but there is abuse, and it was never like that when I started. You've got to have sympathy with the company, because it does cost them a lot of money.

Ian Cummings
There are two thousand working on the assembly line, and how many nationalities do you think? Seventy. Seventy! When

I came here we had English, Irish, Scots and Welsh. The weekend shift – if you go in to the rest rooms during break times you will not hear English being spoken. What a change in society. I'm not saying it's good, bad or anything, but the change is amazing.

Chris Bond

Now with the workforce, you haven't got the comradeship you used to have – we have so many different cultures, so you're never going to have that bonding with everybody. And the amount of women on the plant changed with the equality laws. Traditionally this has always been a male industry, and you've had women in trim shops doing sewing. But now you'll find there's probably as many women working on the line as men. The trade union is very proud.

Elaine Butler

They had a big intake in 1983. The workforce seemed very unstable. They were going out on strikes every other day and a few years later the management thought, 'Perhaps if we have women on the track, perhaps it'll stabilise the workforce.' They'd never had women on the track actually physically building cars. And so I moved to the track, I think it was 2 January 1983. I was putting the steering columns on the Maestro, and at the end we were doing about forty-six cars an hour. I spent six years on permanent nights.

Elaine Butler was born in Oxford in 1953, and began at the car plant when she was fifteen. She trained to be a secretary, and as well as working on the assembly line has since been a personal

*assistant to several managers throughout the factory, including
Paul Chantry. At the beginning, 'It was forms; sometimes I did-
n't even know what I was typing about. When you were in the
offices what they were actually building out there was another
world.' She left for a while to look after her children; her husband Don Butler works at the plant making car bodies.*

When I began here they used to have buses from all around
Oxford. That's where I met him, at the bus stop.

We used to meet up in the canteen at lunchtime and then
travel home together, and one day when I got off the bus he
asked me whether I'd like to go to the cinema that evening and
I said yes.

When I returned on nights I got my children up in the morning for school and was at home when they came home. In the
school holidays they would stay with my parents so it worked
out well. And twenty-five years on I'm still here.

We had a Mini Traveller, red with wood, the only car I can
remember the registration of: PP448. It cost £150 and to fill it
back up with petrol was 50p. We used to go to Bournemouth
with my mum, my dad, myself, my husband and my two boys
in it. And rust. And we had a portable radio in the house that
used to go in the car. Couldn't ever hear anything.

The men I worked with – I cannot fault them. They wouldn't
swear in front of me, they weren't disrespectful. I was just one
of them to them. But when I first came here there weren't even
women's toilets; they'd shut off one section of the men's toilets
and put 'Ladies'. They wouldn't be able to do that now.

I don't even think of it as a male-orientated business because it's just what I'm used to. I think it may be a bit scary for the young girls who are in an office and then come into that environment. But I think ninety-nine per cent of those who've come in have absolutely loved it.

Donna Green

When I joined, the management expectation was that women worked until they got married and then they left. That wasn't my intention, but I did have a case where I was passed over for promotion, and the boss's explanation at that time was, 'Well, how do I know you're not going to get a new boyfriend who's going to whisk you off somewhere?'

Donna Green was born in Cowley, and began working at the plant in 1978. She too remembers the swarm of bicycles at the factory gates, and her father returning home at lunchtime. She says that almost everyone in her road helped make cars, and when she was young she assumed that that was what everyone in England did.

I proved that he was wrong. At the time I accepted the decision but made sure I would focus my work to prove that I was better than the other person.

My grandfather was in what they called the Heavy Gang – they moved bits of kit around the plant. He suffered an industrial accident and lost two of his fingers. That was pre-litigation days.

My dad worked here from a very young age, fourteen or fifteen. He was a millwright, making the tools. In the late seventies, early eighties there was a series of strikes, and because we were a big family and my dad had to work, he crossed the picket line. Suddenly he was completely *persona non grata* with people that he'd worked with for twenty or thirty years, and that later influenced his decision to take redundancy.

Donna Green began as a commercial trainee. She worked in the engineering specifications department, and then the reprographics department, printing drawings for the new Rover Group cars. She got to see the prototypes early. Then it was project-management work at the central headquarters in Canley, Coventry, and by the time the new MINI appeared she was working in quality management and returned to Oxford. Like Elaine Butler, she also met her husband at the plant.

I was given the task of introducing a target management system here. BMW is very target-focused. Targets would be set at the plant level, and then broken down into each technology, so you'd have things like the number of CIP (continuous improvement ideas) which were raised, or targets or limits for absence. It's very much controlled by the central strategy department in Munich and the target cascading process. Of course not everyone is always convinced, and today that is still the case. Today there are different levels of understanding of why these things are important to a company – to know where you are, and where your weaknesses are.

If a group of men are not used to working with women . . .

some of them knew me, not all did, and they were unsure how to behave. Should they use the language that they use every day or not? One of the guys in particular was renowned for using the most appalling language, and yet to this day I've not heard him swear, so they do try to modify their behaviour. And if they do swear during a normal meeting or conversation they will always make a point of apologising to me. Never to anybody else.

Bernard Moss

I'm embarrassed to say I don't know what the percentage of women is but it's higher than any time I can remember. Other than in the trim shop, in all my time in the factory, the chances of seeing a woman before the eighties was a bit like until the sixties you'd have been lucky to see a black man. Our union's always been pushing about ethnics and females and all of that – I'd say, 'Come to Oxford, you'd never criticise us.' The biggest problem we have of course is how many women stewards we have. We've got two, and you've got to have more than that. What do we do? We've got half a dozen ethnic shop stewards, probably more than that if you take in a couple of Slovakians. We've got a couple of Slovak stewards, along with the West Indians, Indian, Pakistani, we've probably got nearly a quarter, so that's good. I think certainly the women bring a bit of life into the tracks, because obviously you get the banter down there. OK, the odd time we get the odd sexual harassment case but not too many fortunately, but that's going to happen. Imagine, there are five thousand people in this plant, so somebody says the wrong thing and we've had a few things like that, but I'm sure they have at other plants.

Elaine Butler

I used to work with Bernard when he was just a shop steward. Out in the palette park, 'Brothers and sisters!' Oh, I used to hate that.

There are loads of women on the track now, loads of them. Obviously to me now the workforce is very young. But some of the women down there, they're absolutely stunning, these Russians, they're absolutely gorgeous and you think, 'These poor blokes, how do they cope?'

The film campaign: the MINI saves the world from invaders

11 'The focus of the launch was the emotionality,' said Frau Dr Larissa Huisgen, twice

Emma Lowndes (MINI marketing director)
They were thinking, 'What does this woman know?'

Paul Chantry
The first thing I remember about the launch was that there was something very different about the way this car was being marketed. We had a visit from a lady called Emma Lowndes, and she was responsible for marketing the MINI in the UK. Very interesting lady. She told us about how the MINI was going to be positioned – it wasn't heritage, but we had to use heritage – all that stuff. It was going to be young, fresh and sexy, so we always used to come to work after that saying how young, fresh and sexy it was to work in this car plant.

She showed us some of the point-of-sale things she had, and one of them was so funny – like, a piece of cardboard and then a frame with cut-out MINIs that you put at the end – it was an environmental positional conceptualised MINI or something. And she said, 'And this enables you, the customer, to visualise the MINI on your drive . . .' And you pulled out the colour you wanted, out of this cardboard thing, and stuck it in there. We were all manufacturing guys, and we walked away from that meeting saying, 'We're doomed.' You know, 'What's going on here? That's the kookiest thing I've ever seen.' And they were going to sell our livelihood on the back of some of those ideas.

And then the UK advertising campaign came out, and it was *War of the Worlds* and 'Mini saves the world. The end.' We realised it was absolutely tremendous and not like anything we'd ever seen before. So different from what we were used to, those boring Rover adverts.

Emma Lowndes
It was important for us to say, 'Don't be like everyone else.' There's a car hierarchy in the UK. You join a company, maybe get a Fiesta, do a bit better you might get the Focus, do a little bit better you might get the Mondeo, and then you make it to the Jag. But in a MINI, just like the old one, you could be the MD or the butcher or royalty – it's just so levelling. So we felt at the time that MINI could be a car for individuals – 'Don't try to put me in a box.'

Emma Lowndes is in a conference room in the BMW head-quarters in Bracknell. In the reception area there is the latest 7-Series, and close by is a MINI customised by the artist Alan Aldridge, an update on the 1965 paint job he did for the cover of the Sunday Times Magazine.

The conference room contains a framed collection of advertising material from dealerships, part of the 'MINI Adventure' campaign to launch the new car in September 2001 ('Couple can't decide which MINI to buy. Sell flat and buy all three. The End'). On the opposite wall are two photographs from the same campaign. The first is from Grey Street, Newcastle, a fibreglass MINI stuck to the side of a scaffolded wall next to the words 'New MINI attempts to drive up building. Succeeds. The End.' The second, of Royal Exchange House, Leeds, shows

Climbing buildings in London (top and bottom) and Liverpool

the same idea, and underneath it the information: 'It's a MINI Adventure. From £10,300 to £11,600 on the road.'

Lowndes had been with BMW for three years before joining MINI in July 2000, a year before launch.

From Rover I inherited a little green plastic file with some PowerPoint presentations in it, a point of view on how you might position the car. It was largely around 'heritage'. I had a flick through it but it just didn't feel right. So I went out and about to talk to some luminaries who knew about brands, just to bounce ideas off them.

I wanted to get a pretty good feel about what British people thought about the old Mini. There was very high awareness: if you said, 'Name a small car,' loads of people would say 'Mini'. Nine out of ten people had sat in a Mini at some point in their life. One thing I realised early on was: to tell people about Mini is quite a hard job, because everyone thinks they know about Mini – part of our national heritage, not just a company or a brand. It's a code: if any newspaper wants a shortcut for 'best of British' they stick in a picture of a Mini. That's a huge responsibility.

I heard through an advertising agency that Peter York, the social commentator, lovely guy . . . I heard it was his area of expertise. I asked him to run a couple of focus groups in London and Manchester. Peter led them, and we had one key insight that came out of that. No one had seen the new car at this stage. We did an exercise in which Peter said, 'Everyone shut your eyes. Think of a Mini experience – you might be driving one, sat in one . . .' And we looked around the table, and everyone was grinning.

Lowndes had also been to see Robin Wight, the chairman of the agency WCRS. WCRS also handled BMW, but a new team was assembled specifically for the MINI.

At the time there was still a lot of animosity about the parting of BMW and Rover, so it was very important in the UK that we demonstrated we understood what Mini-ness was about. Otherwise it would have been everybody saying, 'Oh, it's just a small BMW.' We explored lots of different areas before finally settling on an area that was essentially about its character.

The small car market is the most heavily contested, and at the time of launch if you painted them all silver and lined them all up next to each other it would have been quite difficult to tell them apart. Our job was to show that the new owners, BMW, respected all the achievements of the past – so important, I was fanatical about that, because I didn't want to give the impression that we were trying to own that for ourselves. So it was important to let the new car earn the right to exist on its own four wheels. We just wanted to reinvigorate the brand, and bring it to a new generation to whom the older car was becoming less and less relevant.

The new car was unveiled at the Ecole des Beaux Arts, Paris, at the end of October 2000, towards the end of the Motor Show. A sign at the entrance announced 'Be ready for anything', and there was a party theme before the cars were unveiled – lights, DJs, a speech from the chairman of BMW acknowledging a debt to heritage and a faith in the future.

Three weeks earlier, and two days after the last Mini had

Be ready for anything: Dr Herbert Diess and friend at the Paris Motor Show in 2000

been made at Longbridge, members from Mini clubs in the UK had gathered at the plant in Oxford to see the new car for the first time. 'The car should be judged on its own merits and not against the original Mini,' said Trevor Houghton-Berry, general manager for MINI in the UK. 'If this car has only inherited some of the old Mini's spirit, that is fantastic.' Mini Magazine conducted a vox pop on the new car, and opinions were mixed. 'It's not a Mini, but it is a very nice car,' said Rick from Oldham & District Mini Club. 'I feel very sad about the end of Mini production though.' Lee Pegrum, from Southern Mini Owners Club, said, 'I love it from the front, but am not too sure about the sides.' 'I think the interior's mad,' said Sue from Essex Mini Club. 'There isn't another car out there with one like it.'

Emma Lowndes

That's fine, and for every person who would say 'Oh, it's an abomination,' there would be someone who said, 'I've got an old Mini, I keep it in the garage, and I have a new one for now.' I'd rather work on a car that had a strong image that people either loved or hated than one where people said, 'Oh, it's all right.' It's totally cool – we're not going to force ourselves on anyone who doesn't want to drive our car. That's why we've never given cars to celebrities.

Paul Chantry

For someone like me, who's been involved in lots of launches, you're not convinced of anything until you've seen the sales figures. I always remember the Rover 75. I was on the stand at the Motor Show when Jeremy Clarkson walked in, got in the car and said, 'This is a wonderful car – fantastic!' Then he took it out on the road about a month later and completely slated it.

Autocar, May 2001
**** (out of five)
No new car has been more eagerly awaited in the UK than BMW's interpretation of that great British institution, the Mini.

Reworking one of the 20th century's seminal automotive designs falls only a little short of raising the *Titanic* in the all-time list of difficult tasks. But your first glimpse of the MINI in traffic tells you just about all you need to know. It has been well worth the wait . . . the MINI is as good a piece of retrospective car sculpture as we've seen.

The new car differs most from the original in its packaging, which is nothing like as space-efficient, and is probably best described as a two-plus-two.

With 116 bhp and tipping the scales at a portly 1125 kg, the MINI's sprinting ability slots it into warm-hatch territory. Good traction helps pull the car to 60 mph in 9.3 sec, but accelerating from 0–100 mph takes a lengthy 28.4 sec. A great lugger it is not. Sure, the MINI has enough performance to make it fun, but it's no GTi; at least not in a straight line.

Far more interesting than the engine is the MINI's chassis, especially its ability to involve you in the action. This is a car you aim through corners confident that you're going to clip a blade of grass . . . Above all, the new MINI is a comfortable drive, one whose suspension is as soothing over a rutted road as the first Mini's was crude.

It would be easy to be seduced by the MINI's many charms and award it five stars. After all, it's a great looker that also happens to have one of the best front-drive chassis we've come across. It's great to look at and sit in, but not even the ride and handling or the superb build quality are strong enough to offset the mediocre engine and disappointing packaging. That's a shame, because here's a car every enthusiast naturally wants to like.

Emma Lowndes

I don't want to make marketing look easy, but the clues are all in the product. Always. Robin Wight's attitude is 'Interrogate a product until it gives up its strengths'. Contrary to what people might think, it's about telling the truth.

The car challenged the conventions of the small car market, so we felt that we had to challenge the conventions of small car marketing. We wanted to appeal to people who were keen to try new things, people who might see a movie within the first two weeks, people who might do adventure sports but also think 'I'm going to try painting . . .', people who were inquisitive. So it was about thinking about what media would appeal to them – what would they read, and take time out from their very interesting busy lives to look at. We couldn't just go out and buy lists of these people.

We settled on 'It's a MINI Adventure' in November 2000. In August, September and October we'd gone through a lot of different ad campaigns. One idea was called 'Tiny Elvis'. It was an Elvis character that big [stretches her thumb and forefinger apart] who does things in a MINI, really offbeam. And there was another one which retrospectively was quite Stella Artois-ish. It was Russian peasants digging in a ploughed field having a really terrible time, and then this MINI comes over the horizon . . .

It wasn't all rubbish by a long shot. Another one that was close, or was closer, which now feels very familiar and might have been developed since by another car manufacturer, was one where you'd be doing something, like a football match, and then something would happen that you'd want to distract someone from, and you'd go, 'Oh look, there's a MINI!'

But I think in the end 'MINI Adventure' sort of found us. We thought it could work in a lot of different media, and dealers would like it. We didn't think it would do so well in the vernacular. Bill Bailey did the voiceovers, when he wasn't so well known as he is now. The movie ad was something like 'Annoyingly, Martians invade the earth, but MINI lures them into a trap, The End,' and the MINI goes over a cattle grid, but the Martians with little pointy legs can't cope with it. I still love it so much, an emotional reaction, but the creative birth was not easy. I remember arguing with the director about how many Martians there should be.

Paul Chantry

Peter Morgan was the English project director for MINI who owned it last when it was a Rover project. Peter told me that before the launch the marketing people were getting pretty windy, and they were trying to talk the volumes down.

In Rover we had become used to being invited to the launches. As a manufacturing guy you'd be sat at a table of journalists talking reasonably confidently about your product. With the Rover 200 manufacturing had become respectable, and could be trusted to sit at a table of journalists and not say anything too provocative. But when the MINI came along, none of that happened. The British manufacturing guys were not involved, and there was a feeling of being a bit divorced from the car.

Emma Lowndes

There's a fairly standard marketing textbook which shows normal distribution: you've got innovators at the leading edge, early adopters, early majority, late majority, and your laggers, not a very nice term but just people who wait to see how well something sells before they buy it. We thought that if we're to reach the innovators and early adopters we've got to do low-level things, things especially for them that they could seek out and entertain them. So we did a whole range: Ministry of Sound in Ibiza, running DJs between their different clubs, MINI sandcastles on the beach and towels. Closer to home we did pictures of people doing funny things in MINIs, nothing too rude, not too posed, with the website address written in lipstick somewhere, and we took them to Snappy Snaps and then left them in bars for people to discover. We left matchbooks around. We did guerrilla parking, sticking fibreglass MINIs in interesting places around cities, and car magazines would ask, 'Oh, can we come and sit in it?' while it's on the top of Clapham Picture House or wherever. If you want to be perceived as cool you don't say you're cool, you have to do cool stuff. We did a bondage fetish leather MINI for the London Mardi Gras, and people loved it in the parade, whereas Ford Fiesta would just tie a load of pink balloons to the car and think they'd done their bit.

We broke our films, 'Martians' and 'Zombies', at the cinema. One of the films we did got into the *Guinness Book of Records* as the shortest film ever – a MINI just racing towards you and then away from you in a leafy English lane.

I went to present the campaigns at the factory, and I met the directors, Dr Diess I think, and they said, 'Look, we do our bit, we build a cracking car, but you marketers . . . Rover 75 was a really good car but it just didn't get the support.' So I went in and explained what I was going to do, and half of them were Germans and they were, 'Hmmm, OK, British sense of humour . . .'

Frau Dr Larissa Huisgen

Globally, it was about being different, creating something that the automotive industry had never seen before.

Larissa Huisgen is sitting in the coffee area of the communications and marketing department at MINI headquarters in Munich. Her background was in market research for BMW, but she switched to develop the promotional literature for the MINI.

Outside her office, a MINI sculpture: four Clubman cars are set nose-first in the earth, as if dropped from a great height. There is also a Mini Cooper fixed upside down on an overhead walkway. A sign reads: 'This is "The Other Ranch" inspired by the "Cadillac Ranch" that was built up in Amarillo Texas in 1974 . . .' There is a lot of black space and then the instruction

The Clubman sculpture in Munich

'Please do not climb the vehicles'.
Clearly, the machine no longer belongs to its makers.

If you ask anyone around the world, 'What is MINI for you?' then hopefully there is a bunch of words that should come up immediately: exciting, like the most exciting small car in the world, emotional, design plays a big role, being open-minded, open for new experiences, being tolerant, being global and not excluding anyone.

The MINI team [in Munich] was acting like a start-up. It was part of BMW Group, but it was completely new for the group to have a bunch of crazy people running around being different. Our boss at that time really stressed that we were David against Goliath, and there was a very special feeling.

It all started with CI [communications identity]. I don't know any car company that ever worked with a black background and very intense highlight colours. With that we were really able to make a statement in the market. We also launched the car through the internet, which wasn't so popular at that time. There was a lot about breaking the rules and being cheeky and being small.

The classic Mini brand was very well established, and the fear that we had was that if the new car was just perceived as a small BMW then it wouldn't [succeed]. We did have a discussion about whether we should add the words 'a brand of the BMW Group'. But the decision was not to do it. So everything was done to position the car as a separate brand. The idea was to give it its own values, its own USP, its own character. Also with the dealers, the plan was always not to have MINI mixed with BMW, but to have a separate showroom or corner of a showroom.

MINI Production begins at BMW Group Plant Oxford (Press release, 26 April 2001)
Munich: Volume production of MINI One and MINI Cooper begins today at BMW Group's Oxford plant on schedule for a UK launch in July and a European launch in September.

Around thirty thousand cars are planned for production this year from the plant which has seen a £230 million investment programme in new production facilities for the MINI since July last year.

Commenting on the production launch, Dr Herbert Diess, managing director, said: 'The MINI has met all the stringent BMW pre-production quality targets'. (Notes for editors: See Did You Know factbox attached.)

Dr Herbert Diess (April 2001)
The factory is now fully integrated into the BMW Group international production network. With the start of MINI production the plant begins a new and exciting chapter and takes car production at Oxford into the twenty-first century with a world-beating model – 88 years after William Morris produced his first car in the city.

The most highly utilised plant in Europe: the robots marry the engine to the car

MINI

Larissa Huisgen

The first [European mainland] campaign that we did, the 'Is It Love?' campaign, the aim was to show the relationship between the car and the driver, to the point where the car is perceived as a personality. It was very emotional and young. There was a lot of talk in the meetings about the car having its own personality. The focus of the launch was the emotionality.

The focus of the international launch advertising outside the UK was emotional displacement: people loved their new MINIs more than the things they loved before, including partners, friends, going out to concerts and all previous hobbies. The print adverts featured a picture of the car and: a phone showing a very large number of missed calls; a person throwing their PlayStation in the bin; a Stop sign being used as a pin cushion. Movie advertising featured a drunken man returning to his MINI, caressing the steering wheel and deciding that he loved it so much that he would hail a taxi; a man parking his MINI in a dodgy part of town and lying awake all night worried about its health; and two lads vying for the driver's seat on a twisty mountain road. They all ended with the question, 'Is it love?'

The sentiment echoed a German print campaign from twenty years before called 'Mini. It's like falling in love'. 'Liebe hat sehr viel mit Verwöhnen zu tun. Der Mini sorgt dafur, dass sein Partner nicht zu kurz kommt, den er verwöhnt ihn in jeder Hinsicht!' (Loose translation: Love has got a lot to do with being spoilt. Mini sees that his partner doesn't go short, as he spoils him in every way!)

Paul Chantry
We didn't know we had a success until the sales figures came in.

We started on one shift, and only envisaged going to two, and then the demand was such that we had to introduce the third, to run seven days a week. I think that was a six-month interval from one shift to three, which represented a huge influx of people, a tremendous revolution for the plant. That's when we realised we had a runaway success on our hands. We became the most highly utilised plant in Europe.

For the launch of the Cooper S models, the adverts featured a greyhound looking longingly towards the MINI with a lead in its mouth; an airline steward ushering the viewer into the car as she would a plane passenger; a MINI next to an empty can of spinach; and a MINI eating a Superman cape.

Jeremy Clarkson, *Top Gear* review of Cooper S Works MINI, 2002, to camera, driving around a test track
I've always loved the Mini, the old Cooper particularly. But this new one – I love it even more. I know it's as English as a bratwurst, but it's this interior ambience [wafts his hands], the big speedometer, the funky switches and pedals that are perfectly arranged to do heel-and-toe gear changes – absolutely brilliant.

There are lots of tuning kits available for the new MINI. But this one is from Cooper themselves, and it has the full backing and support of BMW. 0–60 takes 6.7 seconds, the top speed is 140, and best of all, you can feeeeel the extra power, you can revel in the extra grunt, and you can bathe in the noise [noise of grunting engine as he drives].

The thing is, though, does this extra power and oomph upset the standard chassis? Let's turn the traction control off, shall we [looks straight at camera behind him], and find out. [Starts going round bends.] Okay, in medium speed I've just got a very neutral feel, still get that bounce that you get in the standard Cooper, that you used to get in the old Mini as well, very very solid under braking, a bit of under-steer through a steady corner like that, but now full power going out, and you know something amazing – there's absolutely no torque-steer! None at all! Two hundred brake-horsepower through the front wheels and they're coping easily. Just as much fun as a normal MINI [laughs as he goes round corner] but with all this extra power – what more could you possibly want? The only drawback is the price: It's £3,000 more than the Cooper S, making it almost eighteen grand. It's worth it though.

Emma Lowndes
The question now is, 'Can we find a global positioning that works for all of us?' People talk about MINI being an icon, but we will never talk about it being an icon, because we will never rest, we will never give ourselves that accolade.

We challenge ourselves by thinking of ourselves as a style brand rather than a car brand.

Larissa Huisgen
The focus of the launch was the emotionality, but then a year later we said that the car should not be perceived as only a lifestyle brand.

MINI had created a world outside the car, with a lot about lifestyle and values, but now what needed to happen was to focus more on the substance of the car. It's more than a chic design: it's a safe car even though it's a small car; it has everything that you expect from a premium car in terms of engine and safety standards. This of course was important because we know that the classic Mini had some quality issues, and a lot of people still had that in mind when they were asked about MINI, so it was important to say, 'No need to worry if you drive that car – it follows BMW standards,' even though BMW was not mentioned. So the second wave of communications focused a lot more on substance.

Paul Chantry

One started to get carried away with the success of the whole thing. In the second year of production, every month the production plan was recast. From January to December it went up every month. At BMW you make as many cars as you're asked to make – no more and no fewer, because if you make more you're building up inventory and wasting money etc. But by the middle of the year the salespeople were just saying, 'Make as many as you can. If you can make them we can sell them.' In our world that was unheard of.

Whatever you thought of the car a year ago, now it's fantastic. And because the car was so successful, a lot of people came to the plant, and us manufacturing guys had to learn the marketing spiel, and talk about the product values, and if you say it often enough, we all got convinced as well.

The new film adverts featured a passenger opening a MINI door against a lamppost and the lamppost falling over; a MINI pulling a tow-truck rather than the other way around; a woman

moving into a new apartment and seemingly getting everything
she owned into her MINI, including very tall pot plants; and a
passenger hitting the wrong switch on the dashboard and being
sucked out of the car as if by cabin pressure.

Paul Chantry
I remember going to Munich, this was a bit later when the
cabriolet was launched, and I was strolling through the streets
with my wife, and there was a MINI cabriolet outside a shop
with a pair of sunglasses hanging from the mirror. And it was a
sunglasses shop.

Production increased from 42,395 cars in 2001 to 160,037 in
2002. By 2004 it would be about 175,000, aided by the intro-
duction of the Cooper S, cabriolet and high-performance
Cooper Works models. ('You no longer suffer a trudge around
the rev counter waiting for the power to arrive,' Autocar
announced. 'It's now a gloriously frantic affair, goading you
into driving in classic all-out Mini style.') New investment was
needed before capacity edged towards 235,000 in 2006 and
2007.

Frank Stephenson
Somehow it hit the nail on the head. It looks good. It does what
it looks like.

It didn't change much – right through the design the car
pretty much stayed faithful to the original concept. The sketch
from October 1995 with the blue body and the white roof, the
three-quarter front – if you look at that and you look at the
final car, not much has changed.

It brought fun back to motoring. There are enough variables to keep a wide variety of buyers happy. It has sold equally well to female and male buyers. It's a serious toy: some cars come across as not too serious, and other cars as too serious, but this is a serious toy.

HonestJohn.co.uk (on the MINI One, 2001–7)
What's Good:
Great to sit in with retro dials and switchgear, good seats, straight-ahead seating position . . . very desirable indeed in right colour . . . highly recommended . . . the most fun car there is.
What's Bad:
Not much room in the back seat . . . a/c not standard . . . extras to make it desirable can add thousands to the price . . . doesn't look right in all colours: avoid yellow like the plague.
What to Watch Out For:
Do NOT drive through floodwater.

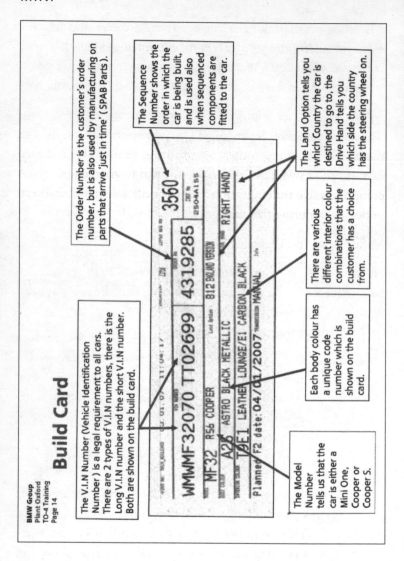

How to make a car: the build card with a customer's options

12 'Kernfertigung integrierende steuerungssystem qualität,' said the manual

MINI People Care (Plant Oxford press release)
The new MINI may be small but it is very sophisticated, in both manufacturing and product specification terms. This required new skills – as well as flexibility – for production associates. A comprehensive training programme for all associates has been undertaken during this past year.

Mike Colley (process trainer)
Hello and welcome to your training day at Plant Oxford, the home of the MINI. Please ask if you've got any questions at any point.

Mike Colley is sitting behind his desk in one of the training rooms built above the assembly lines. It is a large room with a screen for slides and films, and many chairs. Up to eighteen people can attend each session, and several hundred new associates go through the process each year. But on this morning in October 2008 there is only me.

We currently build fifty-three units an hour, and it takes about four and a half hours to assemble a whole MINI from an empty shell after it's painted. It's around four hundred cars per shift. Monday to Friday we operate two shifts, so that's about eight hundred cars per day on weekdays, well over four thousand MINIs per week.

So: health and safety. This is an ESD safety shoe area, electrostatic discharge shoes. Today we'll probably give you these special insert strips that you put in your shoes so that you don't build up static in your body through the day and short out these electrical components.

We don't as a rule wear earplugs in assembly, as most of the tooling is electric, which we call DC tooling. Back in the old days, you'd have these guns which were a lot louder, but now the DC tooling is a lot more quiet and more accurate. We don't wear hard hats, but if you are working under the car and you feel you may be banging your head, you can speak to the manager and he'll get you one.

Gloves. When you go down to assembly you'll see your manager, who we refer to as your PL, your Process Leader, or your TC, your Team Coordinator – they're the guys who will give you some gloves. It used to be PAM, Process Area Manager, but it's changed.

In 2001, when the MINI began, 2,400 associates worked on a single shift to build three hundred cars a day. By 2008 there was also an additional weekday and weekend shift, as 4,700 associates can make more than eight hundred a day. The engines used to come from Brazil but are now made at Hams Hall, near Birmingham, while sub-assemblies are driven down from a plant in Swindon. Manufacture operates on a 'just in time' system, in which the correct parts from about two hundred suppliers are coordinated to arrive precisely when required and in the right order.

We have a Quiet Room, we have many different cultures from many different countries, and that's literally on the other side of this wall should you want to go and pray. We have many different languages in assembly, but while we are training we ask that everyone speaks in English. The reason for that is that neither myself nor the other trainers speak any other languages, and if someone's having a problem and they're talking to their friends in a different language, then obviously we can't understand. Also, every car has a build card on it, and if they can't read that they won't be able to build the car to the correct specification that the customer's ordered.

During your training you will learn a basic understanding of: how to correctly operate DC tooling; how to fit trim parts to the car; how to correctly connect wiring harnesses to electrical components; and the required quality of assembly standards. You will be questioned on the above at the end of your training, a questionnaire consisting of twenty-one questions. We do fail quite a few people, but that's generally down to their poor English.

So, a basic layout of the plant. TO stands for Technology Oxford. So TO1 is logistics, where all the parts get delivered by the suppliers and then sorted. TO2 is body-in-white, where we assemble the body panels which arrive from Swindon. TO3 is the paintshop, where we spray all the cars, and in TO4 we assemble the car, and there's approximately 2,200 people working in here. We have three shifts: Red and Blue, who do two weeks of days and then two weeks of nights, and then the Yellow shift, who work just at weekends. TO5 is quality and engineering, where we feed back all the data, and get feedback from the customers and dealers, and sort out any issues that were occurring in assembly or warranty issues.

The director of assembly is Andy Lambert, this bloke here [a picture comes up on the screen], and he's running the show. But assembly is split into three main areas: there's TA45, that looks after the harness, headliner, cockpit, fuel tank and glazing. There's TA46, which is basically engine, exhaust, seats, bumpers and wheels. And then beneath us it's TA47, test and validation and customer – the final prep, adjusting the headlamps, doing the final checks on the exterior and interior to ensure there's no scratches.

Andy Lambert (director of assembly)
Bringing people into this environment, you don't want to make the mistake that somebody comes in on day one and leaves on day one. You want them to understand what they're signing up for, and if it's not right they make that decision before they start. We can create that environment now where they're far more able to make a sound judgement when they come in. By far the majority go through the training experience, they say yes or no, and those who say yes stay with us for a considerable amount of time. That is very different from the past.

Mike Colley
We should all take pride in building the MINI, and we split this down:

P ride in the Workplace
R espect for the Product
I ndividual Personal Standards
D emonstrate Process Control
E ffective Communication

Ensure that no excess material is left in the car, as this can cause squeaks, rattles and damage, OK? [He puts up another two slides.] These are some scrivets found left in a car. OK, they wouldn't make a lot of noise, however if you've just spent £20,000 on a brand new vehicle, the first thing you're going to do is have a good look around it to see where everything goes and what sort of space you've got. If you lift up the boot and the little panel where the toolkit is, and you find a couple of scrivets there, you're not going to be best pleased.

The emergency stop buttons: there are two types of these in assembly. The Production Stop Button, the Andon button, is for quality issues, and it's on a pillar every ten or fifteen feet down the line, and should only be used by your PL, your TC, or buy-off person. When pressed, this button stops the line and will send a siren to the manager's office indicating that someone wants help, and they will come down the line, see the light and say, 'Andon! Andon!' and you will say, 'Yes, I couldn't get that bolt on,' or whatever. The line will restart when it's pulled back out. The word 'Andon' comes from the Japanese word for lantern – it tells you where the problem is. The Emergency Stop Button is the red button, and can be used by anyone. Once pressed it has to be reset by maintenance, and this can result in a lengthy production stop. It could lead to disciplinary action if either of these buttons is misused.

Ian Cummings
The production line is running with roughly a car every minute – when that process stops, that is serious.

The intensity and immediacy of manufacturing management

is intense. It puts tremendous stress on you. You have thousands of people standing around doing nothing, material flooding into the building, that ain't funny. You have seconds to react to things and make things happen. Your success relies on other people all coming to work at the right time and in the right frame of mind. It is the hardest job I can imagine anyone doing. I know people go on about trawlermen and all that, and I understand that . . . but the intensity . . . there aren't many other things I can think of, in fact none, that are as expensive and complicated to produce in the volumes you're doing them as motor cars.

One manager, Tom Manning, said to me when I began in manufacturing management, 'What on earth are you doing here? I would have thought you would have learnt that there are fantastic jobs to be had in the car industry as long as you have absolutely bugger-all to do with making cars.'

Another manager, ever such a nice guy called Keith Robinson, said, 'Why on earth are you coming this way? Everyone I've ever known in the car industry, if they're in manufacturing, gets out of it.' He asked me, 'Do you like being successful?'

'Doesn't everyone?'

'Well, you're in the wrong place. If you're successful with your volume, you'll be crap at your quality. If you're successful with your volume and quality, you'll be crap at your housekeeping. If you're successful at your volume and quality and housekeeping, you'll be crap at your cost control. You will never win all of it. Ever. You will be criticised for one of the key things, so long as you're prepared for the bollocking that comes . . .'

Andy Lambert

Our aims are the same as any business: how can we achieve higher levels of business with lower cost, and how can we deliver a product that from the customer's perspective has more and more content, again at less cost to us?

We focus primarily on what's the waste in our process. What's the content that no customer is going to be prepared to pay for. If we build the car and make an error, we have to rework that, and that's waste; a customer is only prepared to pay for the single process. There are masses of opportunities still for further improvements. What is the right balance between people and automation to guarantee quality output?

Mike Colley

At the start of every shift and after break and lunch times a buzzer will go, a three-minute warning. If you're not at your station the line will be going, so you need to be there on time. Absenteeism will not be tolerated.

You cannot leave the plant during the early and late breaks, only during lunch breaks. The reason for this is that your first and last break is paid for by the company, and your lunch break is an unpaid break. If you want to go out the gate and over to Tesco or Burger King then yeah, but you have to make sure you're back on time and it's no good saying, 'The queue was massive and I was starving,' because the line will start without you and you will miss five or so cars.

Ian Cummings

We do everything we can in here to reduce variability. Everything we try to do is to work with a process and stick

Once men stood in pits, but now it's the rotary sling

with it. As soon as you vary from it, things start going wrong. The biggest variable in here is people – male, female, age, nationality, mother tongue, you can go on forever. So you train someone to do a certain job, you give them the correct tools, the correct parts and enough time to do that job, and you think, 'Thank god for that – they are unlikely to vary much in what they're doing.' As soon as you move the people around, then you're adding more variability. But you do want to move people around jobs, for health and safety reasons, for repetitive stress injuries, you don't want a guy doing the same thing four hundred times a day for weeks.

You try to give someone a more interesting life on a boring production line, because no matter how much you romanticise MINI and its perception, producing a car on an assembly line, whether it's a MINI or a BMW 7-Series, is still a boring repetitive job.

Bernard Moss

We'll never have an accident-free plant. But with Rover, the accident claims used to go into hundreds and hundreds of thousands each year, and with BMW that's obviously dropped. You've got the rotary slings, where the car is swivelled over so you don't have people in a pit or cakestand like in my day, where the car went up as you stood there. And now you've got all the ergonomic assistors for lifting the doors and petrol tank and batteries. You know, you see a battery and you think, 'What's the problem with that?' But when you've got to do fifty of them an hour, nine hours a day, four times a week . . .

And at one time in the body-in-white, where they weld

everything together, because you had white metal there would be cuts. They've obviously got these leather gauntlets and special coats and all the masks, but a spark can always seem to get in somewhere.

Mike Colley
Right, any questions?

You'll probably find that after your first day your neck or back will be sore where you've been twisting. It normally takes about a week for your body to adjust.

Richard Clay (process trainer)
This is the rear subframe, which obviously holds the back end of the car down. The disc brakes and wheel go on there. This is all lifted on to the car by robot. We make the rear subframe up, and when it's all done the car comes in above it, and a robot comes in to pick the subframe up and screws it into the car with ten bolts. Today in training you will be fitting the lateral arms and the anti-roll bar, which on the assembly line is actually two processes. You have sixty-eight seconds to do your process.

In the practical training room, Richard Clay has begun to demonstrate DC tooling to three potential employees. One of them has come from the paintshop in search of a change of scenery on the assembly line, and one used to work for a rival company making vans. The third has a tape recorder.

DC tooling is a specialised device used to control critical fixings. A critical fixing is classed as a Category A fixing that affects the functionality and/or safety of the car. If any of these

Please do not use the scanners as hammers: the tailgate is welded to a convertible

fixings were not secured correctly it would leave the car unfit to drive or render it immobile. This could lead to serious injury or loss of life and damage to corporate image. And these things could lead on to court cases. It's all bad things.

Critical fixings are measured in newton meters – that's the strength of the fixing. Every fixing may have a different measurement – it could be one and a half or two newton meters, like an airbag crash sensor, or it could be the subframe which could be 150 newton meters.

The way it works is, every car that comes into the unit, the first thing you do is scan it to activate the DC tooling. The tooling is talking to the computer system all the time. We have a system called IPSQ – International Production System Quality, it used to be known as KISSQ [*Kernfertigung integrierende Steuerungssystem Qualität* or Core Manufacture Integrated Quality Management System: every car has a programmable and traceable electronic history, with barcode scanning checking every new stage in the build process]. As the car goes round, DC tooling tells the system that the fixings are correct, and signs it off for that part of the line. At the end of the line, all the processes are stored in the computer, and so if something is wrong they know what to fix in the rework areas.

So the first thing you do on every single car is scan the build sheet which is on every single car. There is one build sheet on the bonnet and one underneath the bonnet. You are scanning the car's exclusive VIN number. Please do not use the scanners as hammers. They cost £400 each and £150 for the batteries. If you're on a process that needs something pushed in, please ask for a little mallet that will be provided to you.

In February 2005, Plant Oxford announced an additional investment of £100 million to enable increased production. But it coincided with the end of the line at Longbridge. Unsold Rover 75s filled huge fields, while new models failed to find investment. A potential collaboration with the Shanghai Automotive Industry Corporation failed. So car production ceased, MG Rover went into administration a few months short of its hundredth birthday and 4,500 people were made redundant. The plant was once the biggest in the world.

The first thing you do after scanning is get a white light on your control box, which means DC tooling is ready to work. You do your fixing, keep your finger on the trigger and it stops by itself. If you get a green light then your fixing is complete and you move on to the next. What you don't want to do then is do it again to check, because you'll get another green light, and DC tooling will be thinking that you've fixed another bolt. So if you've still got three fixings to do, DC tooling will think you've only got two more, and so you'll have a bolt that is not done up. That's when DC tooling becomes a problem – when an operator does double-hitting. If you do double-hit, and it does happen, put a damage sticker on it, and press the Andon button and get your TC down. Make sure you tell somebody – don't just let it go and think, 'Somebody else will find it . . .'

There are many other instructions which appear complex to the novice but second nature to those used to changing their own oil. Always be square on the fixings and work to the principle of Right First Time. Don't let go of the trigger too early, put this bit over the rollbar, locate the stabilisers, sit the bolt here on the lower arm, it's easier if you hold it in to your body, make sure it bites. And then just the same on the other side.

So now we're going to go to the top end of the room and move on to the electrical connections. We will time you on this, but it's not a question of pass or fail. It's to show you that the task has to be performed at a certain speed on the line and you can't just clown around.

You must treat your electrical components with care as they

are easily damaged. Make sure you are not throwing the circuit boards around. If you do drop the components, notify your Team Coordinator and do not fit it into the car as it may cause failure and warranty issues.

In April 2007, less than six years after the start of production, the one millionth MINI rolled off the production line at Plant Oxford. The car had changed a little since its launch – new engines and specifications, a diesel version – but the basic premise was the same: the customer could choose from an enormous range of paint and interior options, so that even though fifty-three MINIs came off the line every hour, it could take several days before the build sheet was identical. (When visitors toured the plant they were presented with the astonishing statistic that it was possible to make one thousand million million cars without making two exactly the same.) Safety and environmental issues also meant that cars built for China, Germany or the United States, or any of the other seventy-five countries where the car was popular, required different statutory requirements. About eighty per cent of production went abroad.

Richard Clay

Do not touch ECT pins, as this can cause static discharge, which can cause failure of the component. When you make a connection, make sure it's in straight and wait until you hear it click into place. The next thing you do is tug the harness just to make sure the connection is made properly. But be sensible about it, just a gentle tug. You'd be surprised the things that have happened in here.

A non-connected connector can cause a rework of half a day

just to find the loose connection, and the whole car has got to be stripped and everything taken out. So if you can't make your connection properly at the start, make sure you tell somebody. And please keep all the lubricants away from the electrics – they do not mix.

There is a desk with sockets and bundles of wires and units marked SPEG and Airbag Control and Audio. The challenge is like something from The Generation Game: *connect everything at its correct location in as short a time as possible. The ideal method involves connecting and then tugging gently to ensure everything is secure, all in one movement. 'It's a flow action,' Richard Clay says.*

As soon as we start I become aware that the others are much faster. Some things just won't go in, or need special techniques I don't have. Rich says things to the other people like 'Good' and 'Oh, that's good'. He says nothing to me. On the tape recording of my efforts I can just hear myself sighing and saying, 'I'm having trouble with this one!'

Then we do the whole thing again and time it. It should take one minute and eight seconds. It takes the others just over two minutes. While I finish, the others chat among themselves: '. . . I've now got a mortgage and kids . . .' It takes me just over eight minutes.

'Eight minutes!' Richard Clay says. 'Not the worst. It once took a guy fourteen minutes.'

After lunch I try it again. The tips of my fingers are sore and chafing. I knock it down to just over five minutes. The cars are being held up all the way down the line.

Peter Crook (director of the paint shop)

There was no automation in the paintshop when I began apart from a large rotary spit. The car would be dipped and rotated in an electro-coat bath to get the primer coat on, and it would then have lead paint manually sprayed. Today it would never meet the legislative rules. The transfer efficiency of the hand-guns was extremely poor, the extraction was extremely poor. The amount of solvent being discharged in the atmosphere was probably ten times what it is now.

The 1 millionth new MINI gets a paint job

Peter Crook began at Plant Oxford when it was only known as Morris's in 1967, and remembers the Mini departing for Longbridge. He worked on the Morris Oxford and the Morris Traveller, while his colleagues built the Austin Cambridge and the Wolseley 16/60. Addressing a framed photo in his office, he points out a bridge that used to take the Marinas to the paint block, now long gone. In those days there would have been only one black, not the three that now stick to his wall in round discs: a mono colour with no metallic just for the roof, one with metallic flake that glistens, one with full metallic.

On the old Mini they used to think that by putting as much paint on as you can, you were protecting the car from corrosion. But that isn't the case. It's all about the quality of the adhesion property, the UV protection, the stone chip resistance.

There are colours that are very very difficult to manufacture. [Pointing to the discs on his wall] These are the three new colours we're introducing: Interchange Yellow, Horizon Blue and Midnight Black. This colour here, the Horizon Blue, has been a nightmare. It went through the lab, two suppliers were approached. One said, 'I can't do that. I haven't got enough knowledge of how to produce that colour.' So we went ahead with the other supplier, and the process is nearly twenty-four months. In the last six to nine months it comes to plant, and having produced the colour, and gone through all the adhesion and corrosion checks and passed all of those, we then need to see whether we can actually apply the colour to the car through our equipment. We found it very difficult – in fact, we didn't achieve it. It's a metallic, and it all depends on the formulation, the pigment content, it's got titan dioxide and we found that it was turning into agglomerates within the paint system. There are so many day-to-day problems with applying paint – you can start off in the morning with one colour, and then the humidity . . .

But you go and look at a current and old Mini and the paint standards – completely and utterly different. On the old one, the rust was there within two to three years.

Once you could choose red, white or blue. Now it's Pepper White, Chili Red, Mellow Yellow, Hot Chocolate, Astro Black, British Racing Green, Lightning Blue, Horizon Blue, Interchange Yellow, Pure Silver, Sparkling Silver, Midnight Black, Nightfire Red, Laser Blue, Dark Silver, Pacific Blue, Rooster Red, Cream White, Dark Grey.

Eddie Cummings
I went back to the plant a few years ago, the first time since I left. Ian took me up to Pressed Steel [now body-in-white] and I couldn't get over the fact that there was nobody there. All you saw were robots, and there were just three or four people in the whole shop.

Ian Cummings
It was ridiculous. We were launching the updated MINI, the R56. Dad was allowed to go into body-in-white, but he wasn't allowed into assembly just in case he saw the updated MINI. As if it would have made a ha'pence-worth of difference – my own dad – how daft can you get? But those were the rules.

13 'Adrenalin is a renewable resource,' wrote the person from the advertising agency

Jim McDowell (vice president, MINI USA)
It's not a sterile piece of sheet metal, it's something special in your life.

Roughly a third of our owners give their MINI a name. When people walk to their refrigerator, they don't necessarily have an emotional reaction. But when people walk out of their house and see their MINI, very often they smile.

Jim McDowell has been in the car industry for twenty-four years. He has worked for Porsche, and was head of marketing for BMW for twelve years before taking over as head of MINI USA. He didn't see a Mini when he was growing up in Colorado, and it wasn't until 1971, on a trip to London as a student, that he was struck by its size and feared slightly for the driver's safety.

We have learnt that about a quarter of the people that buy MINIs will never buy a BMW – at least that's what they've told us. They are BMW-brand rejectors. I think MINI owners are post-materialist. They are buying a car because it is exactly the right choice for them, and if their neighbour doesn't reach the same conclusion, that isn't a problem.

I think everyone takes comfort from the fact that BMW was involved in the design and testing and quality control of the vehicles, but MINI owners march to their own beat.

One of the things that is really unusual is how many people will stop by their MINI dealership to say hello. They're not going there necessarily to transact business. They want to see new cars that have come in recently, they might want to talk to their motoring adviser, they may enjoy saying hello to other MINI owners and showing each other their cars. It's as social a place as Starbucks is. Or Starbucks was in its heyday. When MINI owners take their car for a service, they go into the dealership with the same kind of emotion as when they take their dog to a vet.

But the British history is important to us. We have a Heritage Wall in each of our dealerships, a timeline that starts in the mid-fifties and continues to today, and it looks at how the world has changed in that time and how the Mini has changed.

Larissa Huisgen

It's been about the growing up of the product. When we started off, a lot of people interpreted it as the party and fun brand, but it's just not like that any more. Not just because of the environmental aspects, but the brand has always had a responsibility. That continues to now when we talk about CO_2 emissions and everything under the MINImalism concept.

We started off with one product, which meant that MINI was a car. And then we launched the Convertible and the Clubman, and so what had to happen was to switch from the car equalling the brand to make the brand the roof of a product family. That's still a big challenge – a brand that's sustainable enough to cover more than one product.

Jim McDowell

There are some people for whom their MINI has helped them redefine themselves as a person. Drivers who for whatever reason have said, 'You know, I want to try a different path.' A MINI is a really important step for them, in terms of having a car that is so unusual. Not so much a midlife crisis as redefining themselves. On the female side, this tends to be mothers who no longer have to consider their family first when making a vehicle decision. Previously they have always driven the vehicles that someone else in the family has wanted them to drive. But now they can make a choice for themselves. For guys it tends to be people who are trying to simplify, have fewer possessions, things that are more meaningful but fewer of them – quality things. There are some drivers who are absolutely design-orientated, and for them their MINI is even more important than their home as an expression of their personal identity.

The whole idea of redefining your life works really well with the MINI. People say, 'Why have I been driving a vehicle that's a ton heavier when it's only me inside? Why am I burning up all those resources? Why am I circling the block four times to find a parking space that my gigantic vehicle will fit in?' We had midlife-crisis people at Porsche without a doubt. I wouldn't describe the MINI people as going through a crisis, it's just a realisation that they can do something different than they've always done before.

Larissa Huisgen

We said we are not a trendy brand because trendy always implies short-term. We are not setting trends or following trends, we are a platform for trends. We always want to keep moving and be a step ahead of everyone else. But that's a challenge, because competitors are taking that up rapidly, and we see that in the Fiat 500, really trying to conquer in our territory and they're spending a huge amount of money in their advertising campaigns.

Frau Dr Stefanie Ludorf-Ring (responsible for events, exhibitions and corporate meetings, Munich)
To be honest I think MINI and BMW is not so different any more. Things are very settled here; MINI has now grown up. At the beginning there were no processes, more trial and error, but I think now we are selling so many cars our target is to sell more and more. It's very professional, very process-orientated, and not so creative as a start-up company, which has advantages and disadvantages. The problem is to make people stay: all the people who started with MINI loved the brand and would die for the brand, they were willing to work day and night for the brand, but now they also have to be willing to work within a big company with very structured regulations.

In 2008, BMW placed a viral advert on the internet. For a while this was regarded as the coolest and most effective way of reaching a young audience: you produced a funny or shocking low-resolution film that looked as though it might have been made by keen amateurs, and if it was good, viewers would share it around their friends by either linking to it on websites or sending it by email. It was the personal touch, and it might make the viewer believe that they were not being sold something by a giant German corporation.

The MINI viral was entitled 'Fly', and it was both funny and successful. It was so elaborate and expensive, involving both animation and human actors, that it came with its own 'The making of . . .' film.

The advert opened at a graveyard – a fly's funeral. The fly's relatives and friends are gathered round a grave in distress. An elderly fly leads the service, and speaks in fly language, but there are subtitles: 'We are gathered here today to say farewell to our dear friend Zac.' The flies around the coffin weep. 'We've all lost a wonderful insect. Now let us pay tribute to him, who was so dear to many of us. Between life and death . . . [A fly's mobile phone goes off. The fly looks sheepish before shutting it off.] . . . Between life and death, he chose death, a sensational death. A death that every fly wishes for. [Deeper weeping.] A legendary death. A death of a hero. A death bigger than life! A death that brought heaven closer to earth . . .' The film switches tone. We see a fly hurtling through the air waving a leg as if to say 'Bring it on!' More hurtling, then splat *– the fly has hit something and is dead. The scene pulls out from close-up to reveal the fly on the MINI logo of a moving car, just above the bonnet grille. Another scene change: A MINI rolls across the screen and we read the words 'MINI Clubman. The Other MINI.'*

MINI Clubman sales brochure
Need something different? A new MINI design interpretation. Built for those who love character: the unmistakeable silhouette and unique door design of the MINI Clubman make it instantly recognisable. And it has space for up to five people and their gear.

The Clubman on the drawing board

The options? The interior surfaces and door grips are available in English Oak – where this option is specified, the elliptical door ring is in Piano Black; the Visibility Package ensure drivers can always see clearly – includes a heated windscreen, a rain sensor, a rear-view mirror with automatic anti-dazzle function; the comfort access system allows you to lock and unlock the car without you even having to touch the key – simply having it in your pocket or handbag is enough; the iPod interface allows an iPod or iPhone to be connected, controlled via the MINI joystick or multi-function steering wheel (excluding Shuffle).

The Clubman was launched in autumn 2007, and received a mixed press. Its rectangular end made it an old-fashioned car to look at, and, until driven – and until the options list was consulted – it was the closest the new MINI had ever been to the old Mini. Reviewers agreed that the car looked beautiful, with elegant twin rear doors, and some appreciated the extra four inches of legroom in the back, which still didn't give tall people much relief. The main problem was the new 'Clubdoor', which was an additional half-door to enable rear passengers to get in and out. This was fine on left-hand-drive models, but British passengers found that it swung directly out into the traffic.

Giles Smith, *Guardian*, November 2007
Not for nothing has this portal already been gloomily dubbed 'the suicide door'. And not for nothing have people been chuntering that this supposedly English-blooded car has been built first and foremost for the convenience of the left-hand

drive market, where your passenger door will spend most of its life adjacent to the pavement.

Otherwise . . . it was as punchy and direct as any other new MINI, to the extent that I had to get out a couple of times to check that the new, extruded rear end hadn't fallen off somewhere.

What's more, it was sparky, fun and (until everyone gets one) endearingly different. And let's not forget the practicality. In response to the ancient question, 'How many elephants can you get in a Mini?' the answer is at least two fewer elephants than you can get in a MINI Clubman. Buy one quickly, then, before the estate agents take them all.

Basic price £17,210

Jeremy Clarkson, *Sunday Times,* **December 2007**
I was . . . expecting great things from the Clubman. Because here is a car that offers all of the MINI's edge-thing, Conran-cute design stuff in a package that doesn't force you to amputate your passengers' extremities . . .

Truthfully, I'd rather have a goat.

You can't see out of the back of a MINI Clubman. Glance in the rear-view mirror and all you can see is the pillar where the two doors meet. It's a good job that speeding is now monitored by civil servants in vans, because there's no way you'd see a police car if it were on your tail. And it's a doubly good job because the natural cruising speed of the Clubman S is 110 mph . . . You have to be alert to keep it down, and that's wearing.

But not as wearing as the torque steer. I do not know why the Clubman is so badly affected when the normal car, with exactly the same engine, is not. But I do know that there is no point paying extra for satellite navigation, because this is a car that goes where the camber of the road dictates. You, the man behind the wheel, have no say at all.

Letter to MINI associates from outgoing plant director Oliver Zipse, November 2008
We work in a very dynamic environment where circumstances can change significantly from one day to another. This is why our ability to react swiftly and flexibly to volatile markets remains one of our strongest assets. We have to continue to produce the exact number of cars that have been ordered by our customers.

So what should we do to overcome this difficult time? The answer is: we have to firmly believe in our strengths. I am confident that if we continue to put all our efforts into delivering the best quality and build on our flexible productions system that mirrors actual customer demand, we will weather this storm and come out even stronger.

'Hundreds to lose jobs in MINI layoffs', *Oxford Mail*, 22 November 2008
Almost 300 people are set to lose their jobs at the MINI factory in Cowley, it emerged last night.

Managers at the Cowley car production plant told staff that 290 agency workers would be laid off as the firm attempted to 'better match market needs with production capabilities'.

The move has left the plant's estimated 1,410 agency staff waiting anxiously until Thursday, when they are expecting to hear who will be shown the door. Agency staff account for 30 per cent of the plant's 4,700-strong workforce.

Bernard Moss, union convenor for the plant, said, 'When people are losing their jobs, whether it is one or 100, it's not good. People are going to be concerned – even the ones that won't get tapped on the shoulder will be worried about what happens next year.'

News of the job cuts follows Society of Motor Manufacturer figures released earlier this month which showed 1,886 MINIs were sold in the UK in October – 40 per cent less than the 3,150 sales in the same month last year.

Two weeks ago BMW also announced the factory would close its doors for nearly a month over Christmas – 11 days more than planned.

Peter Crook
So now we're looking at plans up to 2013. With the economic climate at the moment we may be looking at how we achieve a small volume more effectively.

The energy side and the environmental side is quite a challenge to us. The paintshop uses about fifty-seven per cent of the total plant's energy. The cost has doubled in the last fifteen months – my energy bill is something like £8.5 million or £9 million a year. [Since the first full year of production, Plant Oxford claims it has reduced energy consumption and CO_2 emissions by twenty per cent, and water consumption by more than thirty per cent. More than twenty-five different waste materials from production are recycled.]

We've done the basic recycling, but we're looking at deeper projects now. We buy water in, and then, having treated it to an acceptable level, we also pay to dispose of it. But now we're saying, if we do a little bit more treatment on it, can I then reintroduce that water back into our supply? To meet the VOC [Volatile Organic Compounds] regulations, our solvents emissions, we stick those up the stack, and we're limited to what we can stick up the stack. But we're looking at what we can do to limit our VOC even further, to meet legislation that may be coming in 2012 or 2015. To do that we incinerate, burning the solvent before we discharge it into the atmosphere. That means a carbon footprint, but what could we do with that energy that we're using to destroy the solvent – can we recycle that by using the heat from incinerator to heat the buildings?

'What's Your Carfun Footprint?' (US magazine advertisement, autumn 2008)

At MINI, we believe the letters RPM can play nice with the letters MPG. That it's possible to hug trees and corners at the same time. And that adrenalin is a renewable resource. We believe that having fun on the road is not only still possible, it's responsibly attainable.

It is our MINImalism philosophy of doing more with less that has led us to what we like to call the Carfun Footprint. Determined by using a real equation and real math, it is, in fact, a real number. And the 37-MPG MINI Cooper has the best Carfun Footprint on the road.

'MINI's worldwide sales fall', *Oxford Mail*, 6 February 2009,
Andrew Smith

Global sales of the Cowley-built MINI fell sharply in January,
according to the latest figures.

A total of 10,120 cars were sold during the month compared
to 15,457 in January 2007, a drop of 34.5 per cent.

As well as the effects of the recession, bosses said the sales
reduction was due to the lack of a convertible model which
stopped production in the middle of last year with the new
model set to launch on March 28.

The global sales slide mirrors the drop in the number of
MINIs sold in the UK during January which were reported to
be 34.9 per cent down earlier this week.

Meanwhile, plant spokesman Rebecca Baxter said talks were
ongoing between union officials and management over shift
patterns.

Jim McDowell

One of the things I think is really interesting about MINI owners is that they're very optimistic people. They tend to wake up
in the morning with a smile on their face, and they think tomorrow is going to be better than yesterday. We are speaking the
day after the inauguration of a new president. There are a lot of
optimistic people right now. They are not assuming that it will
be an easy path. And it doesn't take very many optimists to
completely sell out all MINIs. So all this discussion about
change could actually be very positive to MINI, even if we're
still in the middle of a recession.

We have a concerning unemployment rate of seven per cent,

but the other ninety-three per cent of the population still have to make choices. And if, say, they're driving a leased Lexus, they may not want to lease a new Lexus and bring a shiny new status symbol to their driveway right when their next-door neighbour is unemployed. But if they came back in a new MINI, that's not the same statement at all. It's more of a compassionate statement. I understand that we have a terrible situation going on in the economy, but people still have leases, and they still have to decide what they're going to do at the end of a lease, and I can see that working in the MINI's favour.

'Global MINI sales up, despite slump', *Oxford Mail*, 9 January 2009

Sales of the Cowley-built MINI plummeted sharply last month as the factory enforced an extended shutdown over the Christmas period.

December's worldwide sales were down 27.8 per cent to 15,010, compared to 20,800 the previous year.

However, annual sales rose 4.3 per cent to 232,425, thanks to a good performance earlier in 2008.

Jim McDowell

Going forward with cars, there will be an array of choice, and the electric car will be on that array. For most people, an internal combustion MINI is just a great solution. But we may find that in our big cities we will increasingly restrict access unless you have a zero-pollution car, and you know what, an electric MINI is a great solution should that future come.

British Pathé News, 26 November 1967
Cleaner and more convenient motoring is the dream of every car owner in Britain. The Commuter, Ford's electric runabout, could well be one of the answers to that dream. In a few years there's the strong prospect of seeing millions of them on the roads.

. . . Over at the British Motor Corporation's Longbridge headquarters they don't intend to be left behind. In conjunction with a battery firm, it plans to build an electric towncar within two years. That's what ace car designer Alec Issigonis is working hard at. [Newsreel shows Issigonis sketching.] This is just a rough idea of what BMC's electric auto may look like, but Britain is way ahead of the world with its plug-in car projects.

The MINI E was launched at the Los Angeles Motor Show on 19 November 2008. It was the usual dry-ice-and-flashing-lights affair, attended by Jim McDowell and BMW CEO Dr Norbert Reithofer, who gave a scale model of the car in a perspex case to Governor of California Arnold Schwarzenegger. The car was built at Oxford, and three days later the staff at the plant had their own launch: a large bonfire party in a field at the back of the visitors' centre. There were burgers and hotdogs, fireworks exploding in synch with music from the James Bond films, and in one corner near a fast-food caravan, the new electric car. Next to this, a promotional video playing on a loop:

Official MINI E film

A high-voltage socket instead of a tank lid, and a battery instead of a tank: the first all-electric MINI is quite different from its conventionally powered cousins. A peek under the bonnet reveals a number of other surprises. Normally the engine would be here, but this car has a huge case over the electric motor which houses all the electronics.

The electric motor, which puts out 204 horsepower, takes the MINI E from zero to one hundred kilometres per hour in 8.5 seconds, and its top speed is limited to 152 kilometres per hour. All the high-voltage components are marked in orange; they can carry up to four hundred volts. The high-voltage

Plug it in at night: the battery pack on the MINI E

components are all completely insulated, so you can open the bonnet and touch anything you want – it's quite safe. Also, the entire car disconnects from the battery when the bonnet is open, so there's no high-voltage electricity anywhere.

The car was in steel grey, with yellow door mirrors and yellow decals showing an electric plug on the sides, roof, bonnet and grille. There was also a decal on the petrol filler cap, which was where you plugged it in. MINI staff stood aside admiringly, and there were comments such as 'No back seats at all!' and 'Four hundred volts!'

Initially the MINI E will be launched as a two-seater, its rear section accommodating the lithium ion battery. The battery will take the car up to 240 kilometres. In the back we have a sixty-litre luggage compartment, with a large cover for the battery. The two air intakes are here . . .

The battery that powers the MINI E is very similar to the type found in a cellphone or notebook. It can be charged from any conventional socket, although the wallbox will reduce the charging time to about two and a half hours.

If anything were to go wrong, or if there was an accident, the first thing you'd do is press the Service Disconnect switch, and turn off the battery. That disconnects the four-hundred-volt unit from the rest of the car.

Once in the cockpit, the driver will have to look quite closely to spot the difference between the MINI E and its conventional counterpart. The MINI E also uses a car key, and has a normal hole to put it in.

All over the plant that week were copies of BMW Group News Special, *the in-house newsletter with a lot of pictures of the new car and some tech spec: the battery is made up of 5,088 cells grouped in forty-eight modules; the car weighs 1,465 kg, with a torque from standstill of 220 nm. Unusually, the battery would last longer in urban traffic than on a smooth motorway, as a generator recovered energy from braking force and feeding it back. Reassuringly for traditionalists, the engine was transversely mounted.*

Almost five hundred people will try out the cars for a year in California, New York and New Jersey, after which they will be returned to BMW for appraisal. The drivers were selected after applying online.

Jim McDowell

It was almost like an adoption process. We wanted to match the car with the families that would have the most success with the electric MINI. They're not going to be driving more than 120 to 150 miles a day, so that rules out some people right there. They have to be structured individuals, because it does involve a certain amount of planning. We've all learnt that we have to brush our teeth before we go to bed. Well, MINI E owners have to learn to plug in their car before they brush their teeth, or they're probably not going to be going anywhere the next morning.

Gert Hildebrand

As a designer, I see myself as someone who must protect the content – you can call it heritage or history or the idea – to protect what is there. The inner values of this car – the smallest possible footprint with the biggest passenger compartment, the general ideas which belong to Issigonis – we have to transfer and protect that for the customer. It's not for us to build ourselves a monument. It's not a Hildebrand MINI. It has to clearly fit the philosophy of 1959.

It's sometimes difficult to protect the simplicity, because marketing and customers and journalists want something new all the time. But the goal is to foster and develop in a Darwinistic

Gert Hildebrand (pointing) with interior designer Marc Girard

way, and to only change a product if it's relevant to the customer. Not for the sake of change. Art is not my game. This is not easy, because in a time like this, where the whole global economy is running down, design is often used as a tool to make dramatic changes to create new demand. But we still have to go along this very positive, evolutionary design.

At the Paris Motor Show in October 2008, Hildebrand and his team unveiled the MINI 4x4 Crossover Concept. It featured four doors, was just over four metres long, and promised four-wheel drive. Internally it had a globe at the centre of the dashboard into which you dropped the key, and a central panel running the length of the car to hold gadgets and refreshments. Later spy photos taken during tests in the snow suggested that some of the more radical design elements had been toned down. The car was expected to go on sale in 2010. It will be the first MINI to be built outside Oxford.

We are generally working five to eight years in advance. The thing about design is that ninety per cent of your ideas are thrown in the bin.

We are not alone in this world – there are two thousand other cars that people can buy and we cannot please everybody. But we must make sure we make something that is good value for money and not a gimmicky trashy product. It's important that you can still drive a Mark 1 MINI when the Mark 3 or Mark 4 is on the road.

Raymond Loewy produced the MAYA principle – Most Advanced Yet Acceptable. A car still has to meet desires, but it

The possible future: a concept from 2008

has to have enough clues and historical genes that you connect it to the predecessor. It's like having a child: you want to see yourself in it, but you don't want it to be a hundred per cent like you.

We work hands-on. You still have to express yourself first on a piece of paper, and of course then all the modern tools are used, clay, computers. We have British modellers here [in Munich] and I believe very much in hand craftsmanship. If you can put it on paper in a simple way, it's the best argument. You can easily judge Yes or No immediately. You cannot be cheated.

Dr Jürgen Hedrich (managing director, Plant Oxford)
The future won't be like today, where we have the Hatchback, the Convertible and the Clubman, and we will always have three MINIs, and all MINIs will always be produced in Oxford. There will be a much broader platform approach.

Jürgen Hedrich became the new plant director at Oxford in January 2009, arriving after four years with BMW in South Africa. He was five when the Mini first went on sale. He is sitting in a room next to the new MINI Convertible in Interchange Yellow, and when he approaches the car he takes the wedding ring from his finger in case he scratches it. The top is down and he looks inside: 'Still not that much room in the back, but a beautiful car.'

It's clear that within the BMW group, MINI will be the expanding brand. But whether it's only one plant, or more, is something

. . . For example, the 4x4 MINI concept goes to Magna Steyr in Austria. How MINI will involve the whole BMW Group strategy – this is the most interesting part. In three, four years' time, when the successor MINI comes, there will be another new ratio . . . and what role Oxford will play in this will be very interesting.

If MINI just stayed two hundred thousand cars a year, with three types, then we would not be developing.

I think driving a car will always be a statement: I'm rich; I'm successful; I care about the environment; I only want to get from A to B. But I think with all this downsizing and talk of CO_2 emissions . . . I think this is exactly the right corner to be in. MINI is a classless brand. MINI just means mini, there's nothing good or bad about it; I drive a small car but I can have a very luxurious one. For me to be MINI was like winning the lottery, the luck after nineteen years in BMW of being exactly in the brand where the growth will be.

OK, it's probably not this year and not next year; as long as people are waiting. They are waiting for money but also technology, people don't know what to buy next – where is the technology going?

In Oxford, the idea of having a plant planned for 150,000 cars and producing 237,700, like last year, now gives us a big advantage. We don't sit on big investment and thousands of people and don't know what to do. Last year we had Saturday and Sunday shifts and a lot of temporary labour, and if we go down a little bit and we're just running to normality with two shifts, then this looks to me like a normal plant. I think the time of automising everything and trying to have a technical solution to everything is a little bit outdated.

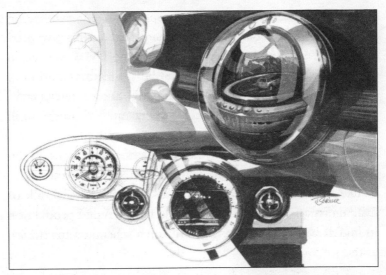

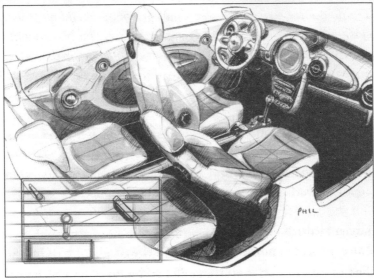

More conceptual possibilities from Munich

MINI

BMW press release, 6 February 2009
MUNICH. Ongoing challenges in the global automotive markets resulted in a decrease in sales at the BMW Group in January. In the month under review the Group sold 70,405 (previous year: 92,849) vehicles of its three automobile brands, BMW, MINI and Rolls-Royce. This was 24.2% fewer than in January 2008.

In the month under review 10,120 (previous year: 15,457, a decline of 34.5%) customers bought a MINI brand vehicle.

The MINI sales reduction was in part due to the lack of availability of the MINI Convertible which ended production in mid 2008 with the new model launch scheduled for March 28th.

Ten days later, worse news at Cowley. The weekend shift was closing with the loss of 850 agency workers. The plant would now only operate a five-day week until demand for the MINI increased. Very early on the morning of 16 February, Ian Cummings informed the associates coming off their last Sunday-night shift – Marzena, Anna, Liam, Yaraslaw, Dritan, Krzystof, Martin, Marta, Dominik, Lucia, Lukasz, Pavel, Jerzy, Apollo, Orazguly, Tomasz, Jakob, Justin, Wajed, Renata, Razim, Richard, Bledar and several hundred others – that they no longer had a job. 'A very very sad day,' Cummings said.

Jürgen Hedrich
The next year or next two years, it will be tough for us. On production, we ran three shifts 24/7, and so reducing just a little bit is just coming back to normality. We don't risk BMW

employee jobs. We talk about casual labour, what a catastrophe it is for the individual, but this is why it is temporary labour. This is a much better situation to be in compared to coming to a plant which ran two shifts nicely through the week and now we're facing a situation of how to reduce thousands of cars. Whether it's a decision, or whether it's luck, this is a comfortable situation to be in.

And we still have world markets, not just one or two, so there are lots of chances. And it [makes us consider] how things will be in five years' time – as I said, there definitely won't be just one MINI plant producing three types. And using this time to prepare for that is probably also good. If you just run and run, then you're not interested in any change and probably you lose focus, or you only focus on volume.

Stefanie Ludorf-Ring

We have three very important things coming up. The fiftieth birthday of MINI, and the main event for that is MINI United at Silverstone for all the drivers and fans. It's a festival activity where people can come and camp, and there are different activities to take part in – different driving experiences; they can watch a MINI Challenge race; different things for families; barbecues; there is a big birthday concert in the evening, then the official launch premiere of two special editions.

The next thing is the international dealer conference, for two thousand dealers worldwide, starting in London ExCel for two days before MINI United. We want to say thank you for everything they have done for us, and what good jobs they've done, and on the other hand to prepare them for the future and for the target to sell more and more MINIs in the next years, and to make them aware of the challenges and how the automotive industry will change.

We will also be looking fifty years in the future, of which one of the most important things is MINImalism. This means, be responsible for the world you live in. The MINI E is part of this, but MINImalism is not only on the product side but also on the political and attitude side. This is what we are planning to do with the brand – to push the brand a little more towards the social responsibility attitude. So it could be that we are going to support a charity or perhaps we also want to make a political statement. We think MINI is the only car brand that can do things like that. We are trying to work out what are the topics of the future where MINI can take responsibility and have an opinion and communicate about.

Larissa Huisgen

We often claim, 'From the original to the Original'. The fiftieth anniversary is very important to underline the fact that the new MINI has transferred the original values of the old.

We are in a challenging time, but I think the products are very good, and I'm very happy that we're not in the first phase of our campaign, when a lot of MINI was about partying and having a good time and no worries. I think this phase is over, and we're now really focusing on all the environmental aspects and value for money aspects, and I hope that will lead us through the difficult times.

Jürgen Hedrich

Tonight I have my first meeting with my guys, talking about

where we want to be in four or five years. That's what I expect from my managers, not to be just on the shop floor every day but to think about new business.

Now is the time to think more creatively and to really respond to ideas that you probably would have called fancy a year ago. I really don't like the word 'crisis', but a little bit of pressure helps us to reflect and see whether we are flexible enough. It's just started, it's not one MINI, it's not three MINIs – it's just started.

Yesterday I saw on television the challenge between English and German people on *Top Gear*. They decided to have a race. Jeremy Clarkson decided to take the British car, the MINI, and then the German lady also took a German car, the MINI. [Once the race was over, Clarkson and the German driver had to leap out of their MINIs and put a towel in national colours on a reclining pool chair.] It was quite interesting. The German lady won.

Gabrielle Hummelbrunner (International Advertising and Media within MINI Brand Management, Munich)
I would like to start with the basic objectives. We would like to communicate the unique heritage of the MINI brand, and its development to a timeless icon. What is very important for all our '50 Years' communication is that we show the past, the present and some aspects of the future. This film will show that the MINI story doesn't end in 2009.

Gabi Hummelbrunner is in a conference room in the MINI building in Munich. She is going through the bullet points of a documentary she is producing to be shown in dealerships and on the internet, to be released in May 2009.

Of course we want to have a new and unseen tonality. We have watched a lot of anniversary films of other brands and have seen that most of them are quite solemn, and we didn't want to have this style, but something very different, something only MINI could do.

We will have two special edition cars for the 50 Years, and these will be part of the film, but the main focus will lie with the brand itself. The world premiere of the film will be at MINI United at Silverstone. It will have pop-culture values.

In the focus of our film will be the MINI drivers and fans, and the story will be narrated by them, so we will show the MINI community. It is very important for us to show that the roots lie in the UK – but now we are a global brand we would like to show all over the world where people drive MINI, so we will shoot in Japan, the US, in Australia and also in the UK of course.

You might know *The Italian Job*, this film, we would like to feature this film in our documentary and do a short interview with [the producer] Michael Deeley and also with Michael Caine. We will show scenes from the old *Italian Job* and also the new one. But we do not want to show only the celebrities, but also the real MINI fan, because it is really amazing the way they put all their heart in the brand, and how they work on their cars. One example is Kasi [a picture of Kasi appears on her screen], and he is one of the wildest MINI collectors and customisers. He got his first MINI at the age of eleven, and now he has over twenty classic Mini and new MINI. He has an amazing garage, but also in his living room you can find Mini, Mini everywhere.

Another example is Olivia Harrison. We will shoot at her villa and make an interview with her, and we will talk of course about the Beatles, the sixties, the classic Mini, all those topics. Another example is the architect Peter Haimerl, and he will talk about the evolution of cities, the design of cities in the future.

In Glasgow there is a MINI driving school, and here are special stunt drivers of MINI . . . Here we have really a very big man with Mini tattoos all over his body – Norm Bullock from Florida. Then we also find out that in Kiev there is a very interesting club scene of MINI. At the end of the film every protagonist will offer their birthday congratulations to MINI in their very own intimate ways.

But we would also like one more thing, to interview people from Oxford – the people who made the car at the very beginning.

Notes and acknowledgements

Because of my very poor speeds on the electrical connections for the airbag control box, the people in charge of the assembly line thought it best that I wasn't let loose on real moving cars. There were customers all over the world waiting for their MINIs, and they didn't want the production held up or things wired up strangely.

But the people who made the car better than I did let me spend as much time with them as I wished, and I am very grateful for their insights and patience. I hope this book is a tribute to their skill.

The book was written with the full cooperation of MINI Plant Oxford, and would have been impossible without the support of the communications department. In particular I would like to thank Rebecca Baxter and John Hawkins for their enthusiasm and faultless organisation. But this is not an officially authorised book, and BMW has had no control over its contents.

I am extremely grateful to the unmatchable Mini archive at the Heritage Motor Centre at Gaydon, Warwickshire, which supplied much archive research material and many of the early photographs. Giles Chapman was extremely helpful in supplying further archive photographs and I am grateful to Suzanne Hodgart for picture guidance. I would also like to thank my editor Julian Loose and all who worked on the book at Faber, particularly Kate Ward for picture editing and design. And I am indebted to Justine Kanter, the only person connected with this book who never owned a Mini, has no Mini stories whatsoever, and still proved ceaselessly loving in all aspects of its creation.

Picture credits and sources

In the text

BMW AG: 123, 128, 136, 137, 139, 217, 250, 259, 262, 264 (top and bottom), 267 (top and bottom)
Giles Chapman Library: 43 (top and bottom), 107, 116, 119, 121 (both), 130 (middle and bottom), 142, 146, 170, 175, 187
The Heritage Motor Centre, Gaydon: v, viii, 5, 10, 12, 20, 23 (top and bottom), 25 (top and bottom), 29, 30, 41 (all), 49 (top and bottom), 59, 68, 72, 73, 77 (all), 81, 83, 93 (top), 94, 97, 98, 100, 103, 104, 105, 109 (top and bottom), 111, 122, 124, 130 (top), 147, 150, 154, 168, 173, 177 (top and bottom), 180, 182, 184, 273
MINI Plant Oxford: 156, 159, 163, 192, 210, 219, 234, 237, 242
MINI UK: 204 (top and bottom), 207 (all)
Roy Davies: 46 (top and bottom), 61, 62 (top and bottom), 63

In the plates

BMW AG: 3, 8, 9, 10 (bottom), 11 (bottom), 12
The Heritage Motor Centre, Gaydon: 2 (top and bottom), 4, 10 (top), 11 (top)
MINI Plant Oxford: 5 (all), 9 (bottom)
MINI UK: 8, 9 (top)

Index

The Heritage Motor Centre

The motor car was invented in the late nineteenth century and has shaped our lives ever since. The collections preserved by the British Motor Industry Heritage Trust at the Heritage Motor Centre in Warwickshire represent the most comprehensive record of Britain's contribution to this revolution to be found anywhere in the world.

The historic car collection features famous marques such as Austin, Morris, MG, Land Rover, Rover, Riley, Triumph and Wolseley and familiar classics like the Mini and Morris Minor. Award-winning exhibitions tell the story of the car and how it was built in more detail as well as showing what goes on under the bonnet of modern cars in 'Under the Skin', an interactive display that will appeal to drivers from eight to 80.

Behind the scenes lies an archive containing business records, sales and technical material, magazines and books, production and engineering records plus a huge collection of images on film, video and negatives. The pictures shown in this book are only a small selection of almost 1 million photographs which combine to provide a record of the motor industry in Britain and the social life of Britain from the Edwardian era to the present.

For more information visit:
www.heritage-motor-centre.co.uk

ff

Faber and Faber – a home for writers

Faber and Faber is one of the great independent publishing houses in London. We were established in 1929 by Geoffrey Faber and our first editor was T. S. Eliot. We are proud to publish prize-winning fiction and non-fiction, as well as an unrivalled list of modern poets and playwrights. Among our list of writers we have five Booker Prize winners and eleven Nobel Laureates, and we continue to seek out the most exciting and innovative writers at work today.

www.faber.co.uk – a home for readers

The Faber website is a place where you will find all the latest news on our writers and events. You can listen to podcasts, preview new books, read specially commissioned articles and access reading guides, as well as entering competitions and enjoying a whole range of offers and exclusives. You can also browse the list of Faber Finds, an exciting new project where reader recommendations are helping to bring a wealth of lost classics back into print using the latest on-demand technology.